The Rarest *of the* Rare

The Rarest
of the Rare

Stories Behind the
Treasures at the
HARVARD MUSEUM
OF NATURAL HISTORY

Introduction by
Edward O. Wilson

Text by Nancy Pick

Photographs by Mark Sloan

HarperResource
An Imprint of HarperCollins*Publishers*

FIRST EDITION

Image used on page i: Carolina parakeet *(Conuropsis carolinensis),* collected before 1811.
Image used on page iv: Male giraffe *(Giraffa camelopardalis),* collected in East Africa in 1902.
Image used on page xii: Male mountain gorilla *(Gorilla beringei beringei),* collected by Harold J. Coolidge Jr. in eastern Congo, 1927.

Book Design by Molly Renda

LIBRARY OF CONGRESS CATALOGING-IN-PUBLICATION DATA HAS BEEN APPLIED FOR.

ISBN 0-06-053718-3
05 06 07 08 ❖ /TP 10 9 8 7 6 5 4 3 2

To my parents, who have made the world a little greener.

—N.P.

To the memory of Sarah Sloan Auman.

—M.S.

melissa parade
(= inyoensis Nabokov
latis auct., not Lot...

argyrognomon (idas)
unnamed s.
Nabokov 19

The Museum of Comparative Zoology

> Struck dumb by love among the walruses
> And whales, the off-white polar bear with stuffing
> Missing, the mastodons like muddy buses,
> I sniff the mothproof air and lack for nothing…
>
> —L. E. Sissman, 1967

Blue butterflies (Lycaenidae), collected in the mid-1900s by Vladimir Nabokov.

Microminerals, collected in the early 1900s by Clarence Bement.

Contents

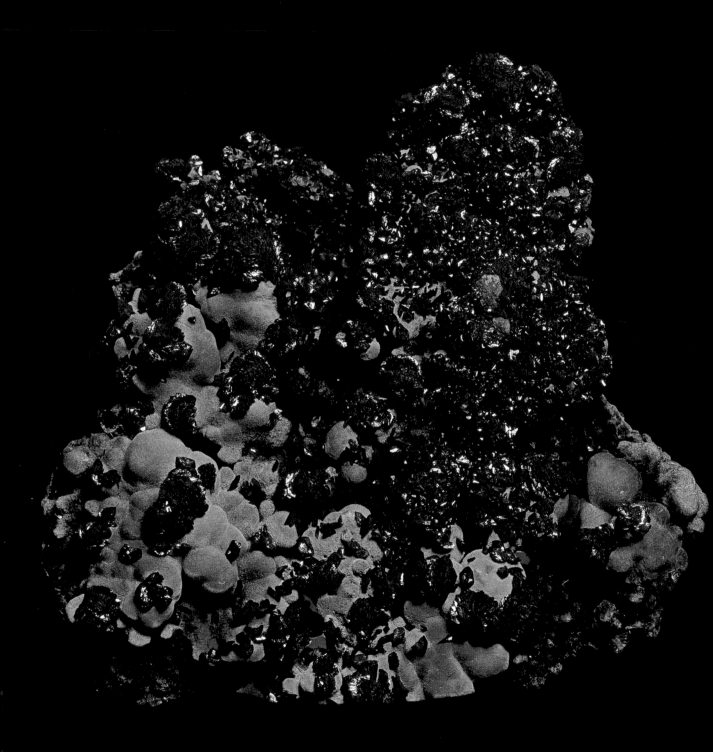

Foreword

A visit to the Harvard Museum of Natural History (HMNH) offers many possibilities. Once past the museum's front doors, visitors follow different paths through the galleries. Some will get no farther than the Glass Flowers, drawn in by petals and pistils, needles and spines, a half-century-long effort to capture the botanical world in glass. Others will stop at the display of endangered and extinct birds, pondering the great auk, the passenger pigeon, the Carolina parakeet, or perhaps the California condor, with its redemptive story of return from the brink. Still others will marvel at meteorites that traversed the solar system before winding up on exhibition, ponder the strange twists of evolution while eyeing the coelacanth, or even vow to reread *Walden* after chancing across the Blanding's turtle collected and donated by Thoreau.

Whatever the path, whatever catches the eye—or the imagination—at the heart of a visit to the HMNH lies the collection, the many thousands of specimens on display and the many millions stored beyond the galleries. Together, these specimens make up one of the nation's great research collections devoted to the natural world. Without them, the museum would inspire neither meaningful scientific activity nor significant public interest. Nor in fact would this book exist, because it is, in essence, a book about the collections.

Edward O. Wilson, in his introduction, puts the collections into a global context. He explains the importance of museums such as Harvard's, which acquire, hold, and study collections, and shows how they contribute to our understanding of nature. Author Nancy Pick, along with photographer Mark Sloan, focus most of their attention (and the body of this book) on the individual specimen. Each bird and beetle, mammal and mineral, fungus and fish, tells a story, and Nancy and Mark have captured a selection of the most interesting. And of course, this represents only the tiniest fraction of what could be told about nature and science, using these collections as a departure point.

There are a variety of ways to read this book. Some who pick it up will be drawn principally to the photographs, struck by the beauty and diversity of the images that Mark has captured through his lens. Some will be lured by the stories Nancy tells of each specimen's history, importance, and intrigue. Others, finally, will find themselves fascinated by the larger picture that takes shape when the images and the stories, arranged in this particular way, begin to create a tapestry revealing nature and the human effort to understand it over time, through the history of a particular institution.

JOSHUA BASSECHES

Executive Director
Harvard Museum of
Natural History

Azurite on malachite, from the A. C. Burrage Collection, collected in Bisbee, Arizona, at the turn of the twentieth century.

The Rarest of the Rare

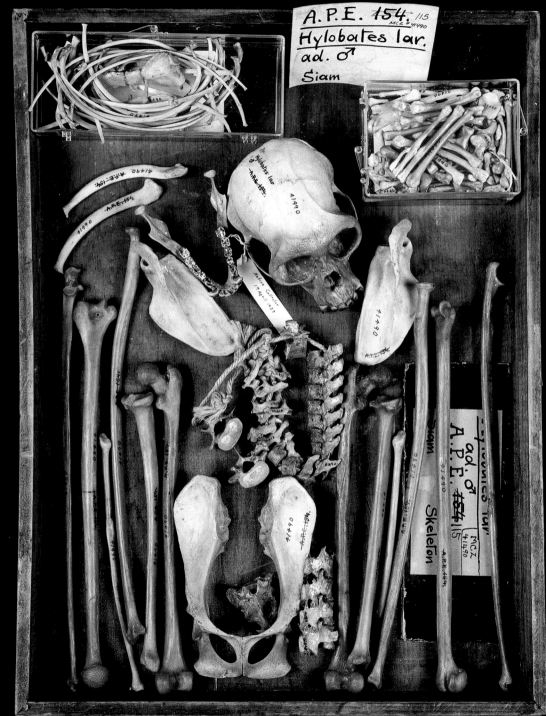

Introduction BY EDWARD O. WILSON

For the slice of time they preserve in human events, we visit battlefields and historical monuments. For panoramas of contemporary life-forms, we travel to zoos, botanical gardens, and wildlife reserves. For knowledge of science and the humanities, we go to libraries and art galleries.

And for all of the above together, we visit natural history museums. At the Harvard Museum of Natural History, which draws on some of the oldest and most richly stocked collections in America, a rock and mineral display combines scientific significance with beauty to chronicle the history of the earth's crust. Close by, selections from the vast collection of fossils span the more than 3.5 billion years in the history of life. The specimens are the actual remains, not pictures or artifacts. They give you a feel of how it would have been to walk through the ancient forests or descend into the oceans lapping the archaic shores. No images, no animations however brilliantly constructed can quite replace this touch of the real thing. They are the closest approach humanity can contrive to an actual time machine.

You may then choose to examine contemporary faunas, with specimens picked by the curators to represent the vast living diversity of life (travel around the world and a dozen ecosystems in an hour!) and the most beautiful and dramatic (see the rare North Atlantic right whale collected at Cape Cod by Alexander Agassiz in 1865!), placed close together for your delectation. In these collections we also

Skeleton of white-handed gibbon *(Hylobates lar)*, collected during Harvard's Asiatic Primate Expedition (APE) in Thailand, 1937.

glimpse life as it was just before modern humans spread around the earth, and we understand what our immediate forebears unwittingly took from us, leaving only the marvelous remains that now stand or lie behind the museum's glass windows. The Harvard Museum of Natural History presents a splendid collection of these: skeletons of a ten-foot moa, a gigantic manatee-like Steller's sea cow, the Tasmanian wolf, an elephant-bird egg the size of a basketball, a reconstructed dodo, stuffed specimens of the great auk, Labrador duck, giant imperial woodpecker and its close relative the ivorybill, and—so easily overlooked—the modest little Bachman's warbler. Other species displayed and on the verge of disappearing might perhaps be pulled back from the grave by the ongoing efforts of conservationists. Among the most important such exhibits of the HMNH are rare specimens of the Javan and Sumatran rhinoceroses, and the whooping crane.

All these displays make for education in science, and the personal fulfillment and capacity for still more pleasure of the kind that grows with learning. Almost every time I pass through exhibitions of the Harvard Museum of Natural History on my way to my office within its bowels, I pause randomly at some place or other and learn something new.

I also absorb the different kind of history centered on this museum. Over the two centuries of their existence, the collections presented by the HMNH, both those open for exhibition or closed off for archiving and research, have been painstakingly built up by generations of collectors and scientists. Their efforts are an important part of the history of biology. The many discoveries they and others in similar institutions have made are now incorporated in textbooks and research protocols, ever valuable but with their origins often forgotten. Within the halls of the museum, what they accomplished comes vividly back to life.

As depicted so well in the chapters to follow by Nancy Pick, the creators and curators of the museum have been almost as varied and interesting as the specimens themselves. Most in the early days were explorer naturalists, an increasingly scarce breed today. They did not have the ordinary interests or lead the pedestrian lives of the generations to which they belonged. Their mission—their obsession, better put—was to find novel things, often in the most remote and forbidding parts of the world. Some of their adventures were the stuff of novels, and their specialized knowledge and focused personalities on occasion beggar fiction. A few were physical and intellectual heroes for the young scientists who succeeded them. Certainly they filled that role for me, as curator of entomology. I live among the spirits of such leaders in my field as Nathan Banks, William Morton Wheeler, and Philip Darlington.

Biology could not have advanced without the collections of museums like this one. Absent their priceless resources, there would be no coherent system of classification, no way to identify the vast majority of organisms, no theory of evolution, no foundation for ecology. Scientists would be traveling through a chaos of kaleidoscopic diversity. Biology would be a slender thread of experimental research reaching up from physics and chemistry and confined to a handful of organisms, probably known variously as "the fly," "the intestinal bacterium," "the cat," "the dog," "the human," "the corn," "the lily," and a few others.

It is widely believed outside biology that exploration of the living world is mostly complete and that the discovery of new species is unusual, so that the role of museums is an all-but-closed chapter of science. Nothing could be further from the truth! Although some 1.5 to 1.8 million species of plants, animals, and microorganisms have been discovered and given a scientific name, experts on classification agree that they represent only a tiny fraction of those actually alive on earth. As many as 10 million remain to be found and studied, and possibly many times that number. Most will certainly prove small, such as insects, roundworms, and bacteria and other microorganisms. But many new species of fish, frogs, reptiles, mammals, and even birds continue to pour in. The bottom line is the realization that we still live on a little-known planet.

Museums and scientists who inhabit them will play a key role, perhaps *the* key role, in the full exploration of life on earth, an effort that is beginning to pick up a new momentum. Museum archives will be the major source of understanding the history of life, and the genetic relationships of the myriad of species, both central goals of biology. From museum specimens, including those preserved in bottles and on pins for centuries, belonging to species both still alive and extinct, DNA and other chemical examples can be taken for analysis. There already exists a long catalog of applications for such data. One of the most unusual is the measurement of toxins in the environment back through history before direct chemical assays could be made. Many such chemicals in the preserved tissues are the same today as when the specimens were first collected.

Collections-based institutions such as the Harvard Museum of Natural History are moving back into the mainstream of science. During my decades at Harvard as student and professor, I have seen two revolutions in general biology. When I first arrived in 1951, Harvard's natural history institutions were still a main center for the university and the outside world. Much of the teaching was focused on particular groups of organisms, following a tradition that dated back to the nineteenth century. Thus we studied invertebrate zoology, entomology, ornithology, malacology, and comparable disciplines devoted to kinds of organisms, in addition to more "general" subjects such as genetics, physiology, biochemistry, histology. A large percent of the faculty were experts on particular groups of organisms. Museums were accordingly still ranked among the premier centers of biological research.

Soon afterward—1953 to be exact—the first revolution began. With the discovery of the structure of DNA that year, and the great impetus it gave to genetics and biomedical research, molecular biology and biochemistry experienced explosive growth. Where biology had been substantially vertical in its structure, divided prominently into expertise on different kinds of organisms, now its orientation rotated ninety degrees to become horizontal, with the emphasis changing to levels of biological organization. Research and teaching focused much more on molecular biology, cell biology, organismic biology, and population biology—the sciences sliced crosswise. The attention paid to comparative studies of different groups of organisms tapered off to a corresponding degree. Museums thus lost much of their cachet as centers of research, and most of evolutionary biology, including especially the discovery and classification of biodiversity, came widely to be regarded as old-fashioned.

By the 1980s, as the study of biodiversity regained prominence, the rotation began to reverse. Museums, zoos, and botanical gardens were newly recognized as centers of research necessary for the progress of biology. Several historical changes caused this turnabout. First, biology was filling out as a complete science and becoming more nearly unified. As the deciphering of genetic codes speeded up, molecular biologists by increasing numbers turned their attention to the diversity and evolution of species. For their part, taxonomists and evolutionary biologists added genetic research to their research programs. Ecology and advanced biogeography, helped by the growing knowledge of biological diversity, advanced toward maturity.

Perhaps most important of all, in this reestablishment of balance in biology, modern conservation science came of age. By the 1990s it was apparent that ecosystems and species are dwindling and vanishing at an ever faster rate, due to habitat destruction, introduction of alien species, and other human activities. The tilting back of biology was not to the vertical that had prevailed before the molecular revolution, but to something in between, say forty-five degrees.

All things considered, great collection-based institutions such as the Harvard Museum of Natural History have much to offer, past, present, and future. Enriched by achievements and history, they are more than ever cabinets of wonder and temples of science.

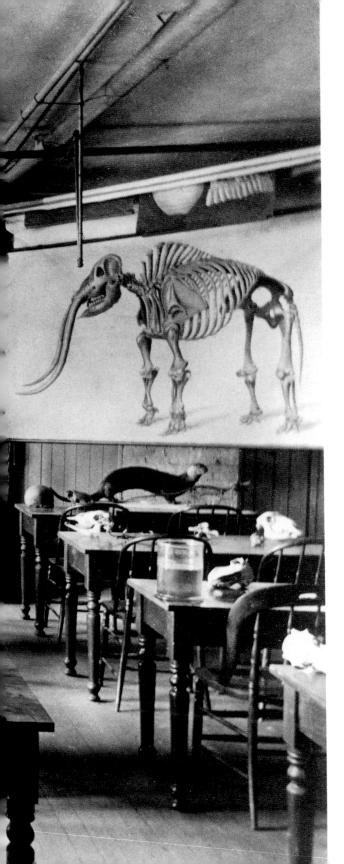

Natural History at Harvard

**In nature's infinite book of secrecy,
A little I can read.**

—William Shakespeare[1]

For anyone drawn to nature's infinite book, Harvard's natural history collections make excellent reading. The collections comprise some 21 million specimens—animal, vegetable, and mineral—from every imaginable part of the earth. Scientists devote lifetimes to their study, deciphering the secrets of species.

Most of Harvard's specimens reside in a sprawling redbrick Victorian edifice, five stories high and not without charm. Housed within is the Harvard Museum of Natural History, which presents the collections—and scientific research based on them—to the public. Permanent exhibitions include the famous Glass Flowers, thousands of animals from around the world, and a beautiful hall of minerals. And yet these galleries offer only a glimpse. For behind the scenes extends a complex of scientists' offices, research laboratories, and specimen storage areas. Past one door lies The Egg Room, with some thirty thousand glass-topped boxes containing birds' nests and eggs. Behind another door reside so many mollusk shells that no

A classroom at the Museum of Comparative Zoology, about 1890.

Harvard's William Schevill (kneeling, at right), hunting for fossils in Queensland, Australia, about 1932. His efforts culminated in the discovery of an immense *Kronosaurus* skeleton.

one has ever managed an exact count. One room yields the world's largest collection of ants; another holds hundreds of horns and tusks; still another reveals meteorites from the asteroid belt. The specimens—most dried, tanned, or pickled in alcohol—fill innumerable bottles, boxes, drawers, and cabinets. They are the world, distilled.

Scientists went to great lengths to obtain the specimens, often traveling to distant shores, amidst adventures and misadventures. The man who collected a rare tailed birdwing butterfly in Papua New Guinea was later eaten by cannibals. The botanist who discovered an exquisite Chinese lily nearly lost his leg, crushed in a rockslide in the remote mountains of Sichuan, the only place where the plant grew. The bones of a huge

Kronosaurus, an ancient marine reptile, had to be dynamited out of Australian limestone by a man nicknamed The Maniac. Other specimens are notable for their collector's fame. Harvard has natural history artifacts from George Washington, Meriwether Lewis, John James Audubon, Henry David Thoreau, Charles Darwin, and Vladimir Nabokov, to name but a few.

Further, Harvard's collections are important for their history. The college was founded back in 1636 by members of the Massachusetts Bay Colony, just sixteen years after the Pilgrims landed at Plymouth. Aiming to turn out well-educated ministers, Harvard taught astronomy almost from the start—fostering discussions of comets, the Copernican system, and discoveries made through early telescopes.[2] Physics came next. After the medical school opened in 1782, the curriculum expanded to include chemistry, mineralogy, and biology. When the great naturalist Louis Agassiz joined the Harvard faculty in 1848, he galvanized the entire country with his vision of an American-based science to rival that of Europe.

Meanwhile, the race was on to discover, describe, and name new species. As intrepid travelers sent back finds from the far reaches of the globe, Harvard's natural history collections grew rich in type specimens—those specimens chosen to represent new species. (Type specimens are considered so important that the British, during World War II, hid theirs in caves.) As Harvard's collections grew, so did the need for better storage and display. During the mid-nineteenth century, scientists shaped what had once been "cabinets" into formal research museums: the Botanical Museum (1858), the Museum of Comparative Zoology (1859), and the Mineralogical Museum (1891).

Gradually, evolutionary science was coming to the forefront. By 1900, Charles Darwin's theory had won wide acceptance by scientists, but many questions about the mechanisms of evolution remained. In the 1940s, Ernst Mayr helped to crack one of Darwin's major unsolved puzzles: How do species originate?[3] During his years at Harvard, Mayr, who became one of the world's leading evolutionary biologists, promoted the usefulness of museum specimens in revealing the process of evolution.

Then came the 1950s, and the discovery of DNA's double helix. The study of museum specimens and "whole organisms" fell out of favor. Molecular biology—genetics—took center stage. The new vanguard dismissed naturalists as stamp collectors and natural history museums as old-fashioned and outmoded. Unexpectedly, with the 1980s came a dramatic turnabout. Led in part by Harvard's Edward O. Wilson, cutting-edge scientists once again began taking interest in the lives of plants and animals. Since that time, natural history museums have become key centers of biodiversity and ecology studies, with biologists searching for ways to slow the current great wave of extinction.

This chapter lays out some highlights of natural history collecting and research at Harvard, over its long history.

Curiosities and Other Early Collections

THE PHILOSOPHICAL APPARATUS. Astronomy came first, becoming part of Harvard's curriculum as early as 1640. Graduating ministers, it was thought, needed to be able to interpret comets and other heavenly phenomena for their communities.[4] In 1672, the college received what was likely its first scientific instrument: an English telescope, three and a half feet long, given by Connecticut colonial governor John Winthrop.

The sciences, at that time, were called natural philosophy. It followed, then, that scientific instruments were known as "the philosophical apparatus." Harvard's apparatus collection grew slowly until 1727, when Thomas Hollis, a wealthy and generous Englishman, donated five chests filled with equipment. Included were instruments for studying mechanics, hydrostatics, pneumatics, and optics.

Hollis also gave money for a professorship in "Mathematicks and Experimental Philosophy," the oldest endowed science professorship in the New World. In 1739, John Winthrop—descendant of the governor—became Hollis professor and Harvard's first important scientist. He ran the colony's first experimental physics laboratory, lecturing on electricity and Newton's laws. Following the terrible Lisbon earthquake of 1755, Winthrop gave two talks on seismology. By explaining earthquakes as natural phenomena, rather than manifestations of divine wrath, he drew the disapproval of more than a few members of the clergy.[5]

The philosophical apparatus, kept in Harvard Hall, continued to grow. The college received an orrery, an instrument that, when turned by a crank, revealed the daily and annual motions of sun, moon, and Earth. In 1748, Francis Archibald donated a human skeleton. In 1758, Massachusetts statesman James Bowdoin gave a valuable microscope. In 1761, science professor John Winthrop set off on an important expedition to Newfoundland, to observe the transit of Venus across the sun. Harvard supplied him with state-of-the-art equipment: an octant, an excellent pendulum clock, and two telescopes—refracting and reflecting.[6] Despite the annoying swarms of insects, Winthrop and his two assistants managed to obtain good measurements.

THE REPOSITORY OF CURIOSITIES. Housed with the apparatus in Harvard Hall was the Repository of Curiosities, Harvard's first collection of natural history and anthropology artifacts. Francis Goelet, a New York snuff merchant, provided the earliest known description of the collection in 1750. In his travel journal, Goelet noted that he visited the "pleasant village" of Cambridge while his ship was undergoing repairs in Boston. He described Harvard's Repository as "not over well stock'd":

> Saw 2 human skelletons, a piece of Neegro's hide tand, &c., hornes & bones of land and sea animals, fishes, skins of diffrent animals stuff'd &c., the skull of a famous Indian warrior, where was also the moddell of the Boston man of warr, of 40 guns, compleatly rig'd &c.[7]

Visitors today would find such a collection offensive, not to mention bizarre and macabre. Goelet's description nonetheless sheds light on a typical natural history collection of the time. Harvard's Repository contained a jumble of human, animal, and historical artifacts, designed as much to surprise visitors as to educate them.[8] America's first scientific natural history collection would not be established until 1784, with the founding of Peale's Museum in Philadelphia.

FIRE IN HARVARD HALL. The earliest scientific artifacts at Harvard have not survived, for on the night of January 24, 1764, amidst the snow and wind of a nor'easter, Harvard Hall burned to the ground. Lost in the fire were telescopes, microscopes, and other scientific instruments, as well as such everyday objects as chafing dishes, tea sets, punch bowls, tobacco, rum, and wigs. Gone were "two compleat Skeletons of different sexes,"[9] as

A view of Harvard in 1743. In 1764, a fire destroyed Harvard Hall, at left, which held the college's earliest scientific collections.

well as the library, containing some five thousand books. The fire also destroyed all the college's collections of interesting objects, including its minerals and the entire Repository of Curiosities, with Harvard's oldest stuffed animals.

The General Court, which had been meeting in Harvard Hall to avoid a smallpox epidemic in Boston, agreed to rebuild the hall at public expense. Kind donors replaced the scientific equipment, mainly with English models, as well as a few instruments made in America.[10] Curiosities and biological specimens also trickled in, including an alligator preserved in spirits, fossil fish, barnacles, moose horns, shells, slabs of Italian marble, marine plants, a bottle containing "curious reptiles from the West Indies," a rattlesnake skin, a catamount (mountain lion) purchased by subscription, a wolf pelt, and a piece of coral given by John Hancock.[11]

A REQUEST FOR SEA HORSE TEETH. On June 14, 1777, a Boston dentist named Daniel Scott wrote to Harvard's governors with an unusual request.[12] His letter is notable as perhaps the oldest surviving correspondence regarding the college's natural history collections. Scott asked that "Two Sea Horse Teeth" be removed from the collections and

granted to him. *Sea horse* was then a common term for walrus.[13] Scott, following the practice of the era, wanted walrus ivory to make dentures.

Scott's letter began with a plea to the humanitarian impulses of the Harvard Corporation, the college's governing body. He acknowledged that a natural history collection had its appeal, namely "the heartfelt Satisfaction of Seeing and Contemplating upon the various Curiosities in the Animal, Vegitable, & other shapes of Beings." But Scott expressed confidence that Harvard would sacrifice the walrus tusks for "the Good of Mankind," given the many benefits of false teeth. During the Revolutionary War, walrus ivory was apparently in short supply.

Harvard agreed to give Scott the tusks, on his promise to provide a replacement set for the "College Museum" as soon as possible.[14] Scott's project may have been ill-conceived, however. According to *The Strange Story of False Teeth*, dentures made

Pheasants that once belonged to George Washington.

You will receive by the Stage the body of my Gold Pheasant, packed up in wool agreeable to your directions. He made his Exit yesterday, which enables me to comply with your request much sooner than I wished to do. I am afraid the others will follow him but too soon, as they all appear to be drooping . . .[15]

I n November of 1786, General George Washington, not yet president, received at his Mount Vernon home an intriguing shipment from France. His friend the Marquis de Lafayette had sent him three Maltese donkeys, a partridge, and seven golden pheasants. According to Lafayette's messenger, the pheasants came from the aviary of Louis XVI, who—still a few years from the guillotine—hunted the specially bred birds in the Bois de Boulogne.

Having learned of the gift, Charles Willson Peale wrote to Washington, saying he was establishing in Philadelphia a museum of "every thing that is curious in this Country." He respectfully requested that Washington send him the pheasants, in the event of their demise. (Peale was already acquainted with Washington, having painted several now famous portraits of him.)

Washington consented. On February 16, 1787, he wrote to Peale:

Eleven days later, Peale replied that the birds had arrived, though the stagecoach was late. "The delay was vexatious," he wrote, "yet I am richly paid in being able to preserve so much beauty."[16] He asked Washington to prepare the remaining pheasants by removing their bowels and putting pepper inside the cavities.

In the meantime, Peale was perfecting his taxidermy techniques. After consulting several European texts, he developed his own system for preserving skins, using first painter's turpentine, then various arsenic solutions.[17] Peale's Museum, the first great American museum of art and science, survived until about 1849. Many of its animal specimens eventually made their way to Harvard—including these two historic pheasants.

from "sea horse" ivory caused bad breath and made food taste disagreeable.[18]

THE CANTANKEROUS MINERALOGIST. In the 1790s, Harvard advanced a step beyond curiosities by establishing its first scientific collection, the Mineralogical Cabinet. This was made possible by John Coakley Lettsom, a prominent Quaker physician and philanthropist in England. In 1793, Lettsom donated to Harvard some seven hundred mineral specimens from all across Europe. The college commissioned an elegant cabinet to display them, eighteen feet long and made of mahogany. Through the glass, students and visitors viewed one of America's first scientific collections of minerals, with specimens systematically arranged and documented. Among the many visitors were Daniel Webster and George Washington.

Lettsom's main tie to Harvard was his friend Benjamin Waterhouse, a professor in the newly founded medical school. In 1788, Waterhouse also began teaching at the college, offering a course on natural history. His early lecture topics appear somewhat odd to modern eyes, ranging from Linnaean classification to "vegetables," "bees and honey," "the terraqueous globe," "humanity to brutes," and "the important process of digestion."[19] Gradually, mineralogy became his main focus, as Harvard's collection grew and Waterhouse himself learned more about the science. By the late 1790s, Waterhouse's natural history course had become quite popular with students, reaching enrollments of more than sixty, a considerable proportion of Harvard's student body.[20]

Waterhouse's lasting fame, however, rests not on his pioneering efforts in mineralogy but on a quite different accomplishment. In 1800, he became the first physician in the United States to test the vaccine for smallpox. His first subject was his own five-year-old son. Waterhouse inoculated him with a cowpox sample sent by Dr. John Haygarth, England's leading authority on infectious diseases. Following further testing, in which vaccinated patients were deliberately exposed to smallpox, the vaccine was accepted throughout the United States.

Waterhouse left Harvard in 1812, shortly after losing his professorship due to disputes with the college's medical establishment. He was by all accounts a cantankerous and pedantic man, disliked by many. As for Lettsom, his role as "patron of American science"[21] has been nearly forgotten. What keeps his name alive is instead a humorous verse, concerning his alleged penchant for bleeding his patients. It survives in many versions, this one perhaps the best:

> When any sick to me apply,
> I physics, bleeds, and sweats 'em;
> If, after that, they choose to die,
> Why, Verily! I Lettsom.[22]

THE MASTODON MURDER. One Harvard mineralogist whose name has gone down in history is John White Webster. Like Waterhouse, Webster was both physician and scientist. He brought new coherence to the Harvard Hall mineral collection, arranging specimens by chemical composition,

The Harvard mastodon in 1874, shown in the Boylston Hall natural history room. The mastodon was later moved to the Museum of Comparative Zoology.

external characteristics, geological characteristics, and geography.

Unfortunately, Webster's personal life was not so well ordered. He was known for living beyond his means, supporting four daughters and a habit of lavish entertaining. Even his purchases for the college outstripped his financial wherewithal. In 1846, he set his sights on acquiring a rare, intact mastodon for Harvard's natural history collection. Attempting to meet the $3,000 price tag, he obtained pledges from private subscribers. When several failed to pay up, Webster was left scrambling. For help, he turned to George Parkman, a wealthy physician, landlord, and fellow Harvard graduate.[23] Parkman sent around a collector on Webster's behalf, to pressure wayward mastodon supporters.

Webster was already indebted to Parkman, having borrowed several hundred dollars from him a few years earlier. By 1847, Webster's financial straits had become truly dire. He borrowed an additonal $2,000 from a group headed by Parkman. As collateral, Webster put up all his personal property, including his prize possession—his private collection of minerals. Parkman was irate when he discovered, the following year,

The building that houses Harvard's natural history collections, shown in 1911.

that Webster had tried to sell the very same mineral collection to Parkman's own brother-in-law for $1,200. Parkman grew adamant about getting his loan repaid, hounding Webster on the street, outside his home, and even in the classroom, where he would sit in the front row and glare.[24]

On Friday, November 23, 1849, Parkman paid a visit to Webster at his Harvard Medical School office—and then disappeared. It took a week before investigators, aided by a suspicious medical school janitor, discovered Parkman's dissected body. They found some parts hidden in a chamber below Webster's laboratory, others burned in his furnace, and still others in his tea chest. Parkman's remains were so mangled that they could not be positively identified, though his distinctive false teeth provided key evidence.

The trial was sensational. People were aghast at the thought that a respectable Harvard professor might commit murder. A record sixty thousand people came to watch, each admitted for a ten-minute peek. In a precedent-setting case, establishing the power of circumstantial evidence, Webster was convicted and hanged.

In its time, this was "the most notorious murder in American history," wrote criminal law professor Alan Dershowitz, in his introduction to the trial transcript reprint. Even today, he noted, lawyers continue to debate Webster's guilt or innocence. The case has inspired a film by Eric Stange, *Murder at Harvard*, and a fascinating quasi-historical account by Simon Schama in his book *Dead Certainties*.

The Making of the Great American University Museum

AGASSIZ AS VISIONARY—AND CREATIONIST. Indisputably, Harvard's renown as a center of scientific learning can be traced to Louis Agassiz (LEW-ee AG-a-see). Agassiz grew up in Switzerland, the son of a minister, and moved to Paris in 1831 to work with Georges Cuvier, then the greatest naturalist in Europe. As a young man, Agassiz authored a groundbreaking study of fossil fish and pioneered the then controversial concept of a great "ice age."[25]

Invited to teach at Harvard, Agassiz arrived in 1848 with a grand vision: to found a natural history museum in Cambridge that would rival the museums of London and Paris. This was

Louis Agassiz at the chalkboard in 1872.

The year 1859 marked the opening of the partially completed museum, as well as the publication of Charles Darwin's *Origin of Species*. The irony here is trenchant, for Agassiz's vision of the museum excluded all notions of evolution. To his dying day in 1873, Agassiz believed that species did not change over time. He held to "special creation," the idea that successive populations of animals had been created and destroyed by an intelligent Creator. He wanted the Museum of Comparative Zoology to demonstrate God's master plan for the animals, emphasizing the individual creation of every single species.[27] He intended the arrangement of museum specimens to reveal the designs of a thinking deity, showing similarities of form, development, and geographic distribution.[28] As Agassiz wrote in a report to the Massachusetts legislature in 1868, "If I mistake not, the great object of our museums should be to exhibit the whole animal kingdom as a manifestation of the Supreme Intellect."

A LESSON IN LOOKING. Evolution did not win instant approval among scientists, but rather gained acceptance over the decade of the 1860s. Agassiz continued to attract excellent students, and he grew famous for teaching the art of observation. "Study nature, not books" is the dictum he is known for (though he likely did not state it exactly so). When a new graduate student arrived at his laboratory, Agassiz would hand him a specimen and instruct him to spend several days looking at

considered a preposterous idea at the time, and yet Agassiz, a man of enormous charisma, made considerable progress toward his goal. Oliver Wendell Holmes, the essayist and Harvard Medical School dean, described him as "robust, sanguine, animated, full of talk, boy-like in his laughter."[26] In 1859, Agassiz convinced a tightfisted Massachusetts legislature to contribute the then impressive sum of $100,000 toward the future Museum of Comparative Zoology. The legislature, in return, required that the museum be open to the public, as it remains today, through its partnership with the Harvard Museum of Natural History.

Before Carl Linnaeus, scientific naming was in a muddle. Naturalists often gave long and unwieldy Latin descriptions to plants and animals, then changed them at will. Linnaeus did not invent the system of binomial nomenclature—the two-part Latin name—but he did refine and popularize it. The beauty of Linnaeus's system is its simplicity: every life-form receives a genus name and species name, and these remain mostly stable over time.

Scientific naming took a major leap forward in 1753, when Linnaeus published his two-volume *Species Plantarum*, describing many species of plants. Botanists derive the first official plant names from this book. For zoologists, the starting point for naming came a few years later, when Linnaeus published the tenth edition of his *Systema Naturae* in 1758. This huge compendium provided brief descriptions of all the plants and animals known at the time, assigning each a two-part name. Many of these names are still used today.

In the century after Linnaeus, there was huge growth in the number of species being identified. Scientists and adventurers, reaching distant parts of the planet, raced to be the first to name and describe new organisms. This mission had an imperialist side, to be sure, as many plants and animals already had perfectly good local names. But scientific names carried the advantage of universality. The Linnaean system eliminated the confusion of having, for example, a butterfly called the mourning cloak in the United States, the yellow edge in Canada, and the Camberwell beauty in Britain.[29] People all over the world, whatever their language, can understand *Nymphalis antiopa*.

Scientists and naturalists sometimes disagree over names. If two scientists propose names for the same species, the valid name (barring technicalities) is the one that first appeared in print. This is known as the principle of priority. Occasionally, scientific names change over time, as scientists reclassify species in light of new research. A revised name may remain controversial, as no central agency makes rulings in the case of a dispute. Generally, scientists cite the authority with whom they agree and strive to develop consensus among scholars.

More recently, some biologists have advocated overthrowing the entire Linnaean system. In its place, they would use something called the PhyloCode, a new naming system that grew out of a Harvard workshop in 1998. The PhyloCode, based on the evolutionary history of organisms, assigns names to groups that share common ancestors.

This book retains the standard Linnaean practice, presenting scientific names in italics, with genus name followed by species name. The endnotes also list an additional name, in standard type, that of the person (or persons) who named and described the species for science. For quite a few of this book's specimens, the describer is none other than Linnaeus himself.

it, without so much as a magnifying glass. "Look, look, look," he would insist. One former student, Samuel H. Scudder, described the experience in an 1874 article. Scudder spent an entire morning staring at his fish specimen, a grunt, finding it increasingly loathsome. In a moment of desperation, he recalled, "I pushed my finger down its throat to feel how sharp the teeth were. I began to count the scales in the different rows, until I was convinced that that was nonsense. At last a happy thought struck me—I would draw the fish; and now with surprise I began to discover new features in the creature."[30]

Scudder couldn't sleep for thinking about his fish, trying to come up with observations that would satisfy Agassiz. But whenever Scudder inquired what he should do next, Agassiz gave the invariable response: "Oh, look at your fish!" Scudder, who went on to become a distinguished expert on fossil insects at Harvard, wrote that this lesson in observation was Agassiz's great legacy, a lesson of "inestimable value."

Agassiz was no saint, and he held racist views that have today tarnished his name. Yet it would be difficult to overstate his centrality to American culture during his lifetime. He was a familiar name in many American households. He socialized with Ralph Waldo Emerson and Henry Wadsworth Longfellow.[31] His second wife, Elizabeth Cabot Cary Agassiz, went on to become founder and first president of Radcliffe College. Henry Adams, in his classic memoir, *The Education of Henry Adams*, wrote that Agassiz's lectures on geology and paleontology at Harvard had "more influence on [my] curiosity than the rest of the college instruction altogether."[32]

And Agassiz's contributions to the future of science were also significant. He left the Museum of Comparative Zoology with huge collections, having inspired natural history enthusiasts across the world to send him specimens, particularly the turtles and fish that were his specialty. He trained an entire generation of distinguished naturalists, including the founders of the Marine Biological Laboratory at Woods Hole and the American Museum of Natural History in New York. If Cambridge had once been considered something of a scientific backwater, Agassiz put it on the map.

ASA GRAY, DARWIN'S FRIEND. Agassiz, for all his charisma and influence, became increasingly intellectually isolated during the 1860s, as evolution gradually became accepted by the majority of scientists. To make matters worse, he regularly had to endure contact at Harvard with Asa Gray, the great American champion of Charles Darwin's theories.

Gray, a noted botanist, had come to Harvard in 1842. He had first met Darwin in London a few years before, when both were in their twenties, and the two began to correspond regularly. Indeed, Gray received one of Darwin's most famous letters, dated September 5, 1857. "My dear Gray," it began, "I forget the exact words which I used in my former letter, but I dare say I said that I thought you would utterly despise me when I told you what views I had arrived at . . ." The letter then laid out a summary of his theory of evolution by means of natural selection. Darwin wrote that even his oldest friends had attacked his ideas, and that he had begun expecting his views to be everywhere "received with contempt."

organic beings at all times. These always seem to branch and sub-branch like a tree from a common trunk; the flourishing twigs destroying the less vigorous, — the dead and lost branches rudely representing extinct genera and families.

This sketch is most imperfect; but in so short a space I cannot make it better. Your imagination must fill up many wide blanks. — Without some reflexion it will appear all rubbish; perhaps it will appear so after reflexion. —

                  C. D.

This little abstract touches only on the accumulative power of natural selection, which I look at as by far the most important element in the production of new forms. The laws governing the incipient or primordial variation (unimportant except as the groundwork for selection to act on, in which respect it is all important) I shall discuss under several heads, but I can come, as you will easily believe, only to very partial & imperfect conclusions. —

The last page of Charles Darwin's historic 1857 letter to Asa Gray.

A year later, Darwin rushed excerpts of this letter into print, to prove that he had already elaborated his theory. The reason for the haste was that Darwin had received a paper from the British naturalist Alfred Russel Wallace containing ideas shockingly similar to his own. Darwin's influential friends quickly arranged for the publishing of a joint paper with Wallace in 1858, in the journal of the Linnean Society of London. (Wallace, collecting insects on the other side of the globe, was not even told about the arrangement.) Ultimately, of course, Darwin received nearly all the credit for the theory of natural selection. As for the seminal

letter to Asa Gray, written in Darwin's crabbed script, it remains in the archives of Harvard's Botany Libraries.

Meanwhile, relations between Gray and Agassiz, the antievolutionist, were deteriorating. The two debated at the Cambridge Scientific Club in May 1859, when Gray introduced Darwin's theories for the first time. Gray noted that he wanted "partly, I confess, maliciously to vex the soul of Agassiz with views so diametrically opposed to all his pet notions."[33] In 1863, the two broke off relations, apparently in the wake of an ugly argument on a train. They had no contact until 1866,

when Agassiz offered Gray a "complete and satisfactory apology,"[34] and they then resumed polite discourse.

Beyond his role as Darwin supporter, Gray was renowned for his textbooks and for *Gray's Manual of Botany*, a plant identification guide. He published widely on the plants of the United States, and he named and described many species collected on the major expeditions of the nineteenth century. During his career, he also assembled important collections of preserved plants. In 1858, he wrote to Sir William Hooker, director of the Royal Botanic Gardens at Kew near London, saying he intended to establish a botany museum at Harvard "in humble imitation of Kew."[35] Hooker kindly sent Gray duplicate specimens of pods, cones, nuts, wood samples, and other commercially important plant products. With these as its core collection, Harvard's Botanical Museum was officially founded, under its original name, the Museum of Vegetable Products.

At the same time, Gray continued to enlarge his personal herbarium—a collection of dried, pressed plants—which he gave to Harvard in 1864. This, named the Gray Herbarium, became the principal center for botanical research in the United States in the latter part of the 1800s. Harvard's various plant collections now come under the umbrella of the Harvard University Herbaria, containing some

5 million specimens of dried plants and fungi. They represent an astounding range of life-forms, from microscopic diatoms to giant sequoias.

ALEXANDER AGASSIZ, PHILANTHROPIST AND OCEANOGRAPHER. Alexander Agassiz succeeded in one area where his father, Louis Agassiz, did not: he made money. A fortune, in fact. In the late 1860s, he salvaged his brother-in-law's unproductive copper mine on the Upper Peninsula of Michigan. Within a few years, Alexander Agassiz had become president of the Calumet and Hecla Mining Company, which became immensely profitable. Over the years, he would donate more than $1 million to the Museum of Comparative Zoology. Through Alexander Agassiz's gifts and influence, several wings were added to the original 1859 building. These partly fulfilled Louis Agassiz's vision of a large quadrangle that would house departments for the various branches of natural history, including animals, minerals, and plants.

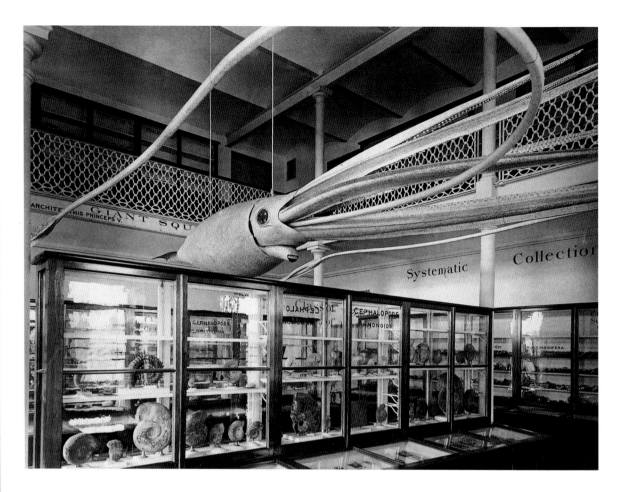

Papier-mâché model of a giant squid, made about 1883 and photographed at the Museum of Comparative Zoology in 1892.

Alexander Agassiz also undertook important scientific work, primarily in oceanography. With naval commander Charles Sigsbee, he pioneered the use of steel cable for dredging. This represented a vast improvement over hemp rope, which often snapped while being hauled in. In the days before deep-sea submersibles, dredging was the only way to obtain biological specimens from the seafloor, and the method is still used today. Agassiz became an expert on echinoderms—a group including sea urchins and sea stars—and published more than 150 articles and books.

Unlike his father, Alexander Agassiz quietly accepted Darwin's evolutionary theories. He did, however, wage his own battle against Darwin, on the subject of coral reefs. Darwin believed that atolls—ringlike coral islands surrounding a lagoon—had been formed by sinking volcanoes. Agassiz dismissed Darwin's sweeping theory and instead attempted to explain the formation of each atoll based on local conditions. Obsessed by his quest to discredit Darwin's concept, Agassiz visited virtually all the coral reef regions of the world, at huge expense. He led expeditions to Hawaii, Fiji, the Great Barrier Reef of Australia, the Maldives, and other destinations. And yet, like his father before him, Agassiz lost the debate. Scientists have since proven that Darwin's theory of atoll formation is essentially correct.

Alexander Agassiz deserves the credit for transforming the exhibitions at the Museum of Comparative Zoology, making them quite sophisticated for their time.[36] Taking over as director at his father's death in 1873, Alexander Agassiz oversaw the museum for nearly forty years. He based his plans partly on his father's intentions, arranging displays according to geography and "synopsis"—typical examples of animal families. There were rooms devoted to the fauna of Africa, India, North and South America, Europe, and Australia. Other galleries displayed fossils, arranged by geologic period.

A VISIT FROM ALFRED RUSSEL WALLACE. These exhibitions greatly impressed Alfred Russel Wallace, the English naturalist who, independent of Darwin, had posited a theory of natural selection. In 1886, during a lecture tour of the United States, Wallace made two visits to the Museum of Comparative Zoology. He then published a glowing account of the museum in a popular English magazine, *The Fortnightly Review*, applauding its organization and presentation. Harvard's museum was, he wrote, "far in advance" of the British Museum, which he described as displaying "countless masses of unorganized specimens"[37] in gloomy halls.

The British Museum had recently moved its natural history collections into a breathtaking new building in South Kensington, London. And yet,

wrote Wallace, "the great bulk of the collection still consists of the old specimens exhibited in the old way, in an interminable series of over-crowded wall-cases, while all attempt at any effective presentation of the various aspects and problems of natural history, as now understood, is as far off as ever."

By contrast, Wallace found the Museum of Comparative Zoology's public galleries thoughtfully arranged to avoid overwhelming the visitor. All specimens were displayed in glass cases, with neat labels. The vast majority of the collections, then as now, were kept in storage areas out of public view. Wallace found the museum's geographic organization—still used in some of the galleries today—highly instructive.

THE GLASS FLOWERS. Wallace might have enjoyed his visit all the more had he waited another few years, for Harvard had yet to commission its best-known natural history display. The first Glass Flowers arrived in Cambridge in 1886, broken. Yet to George L. Goodale, they looked beautiful. Goodale, director of Harvard's Botanical Museum, had long been dissatisfied with the crude botanical models available at the time, made of wax or papier-mâché. He wanted superior models for his students and visitors, and glass appeared to be the answer.

The models had been fashioned by two men, Leopold Blaschka (BLOSH-kah) and his son, Rudolf, in their studio near Dresden, Germany. For years, the Blaschkas had enjoyed considerable success making glass models of jellyfish, sea anemones, and other marine invertebrates, much sought after by museums worldwide. Goodale traveled all the way to Germany to meet with the Blaschkas, hoping he could persuade them to switch to creating botanical models. Eventually, with financial backing from the Ware family, he convinced the Blaschkas to sign an exclusive contract with Harvard—perhaps little suspecting that the Glass Flowers project would take fifty years to complete.

All told, the Blaschkas created more than four thousand botanical models for Harvard, from elegant irises to diseased apples. The models—officially known as the Ware Collection of Blaschka Glass Models of Plants—are so realistic and detailed that visitors often ask, "Are they really made of glass?" Truly, the artists were nothing short of obsessive in their quest to replicate real plants. One professor viewing the collection, himself obsessed, counted twenty-five hundred buds and flowers on the model of the angelica tree.

The Glass Flowers continue to have scientific value, as accurate models of real plants all in bloom at the same time. But perhaps more importantly, they have endured as exquisite monuments to the meeting of art and science. They have been featured in countless articles, as well as a Marianne Moore poem ("Silence") and a Jane Langton mystery novel (*The Shortest Day*). Avant-garde photographer Christopher Williams used images of twenty-seven Glass Flowers, classified by the plant's country of origin, to protest political disappearances.

WEALTHY MEN'S MINERALS. In the meantime, Harvard's Mineralogy Cabinet was poised to enter the modern age. In 1850, following the hanging of John White Webster for murder, Josiah P. Cooke Jr. took over as curator of the mineral collections. Vexed to find the collections in disarray, Cooke took the humbling step of inviting a Yale mineralogist to sort things out. Once the worthless pieces had been culled, Cooke spent several decades acquiring fine specimens through purchase, gift, and exchange. In 1875, the collection was recataloged according to the latest scientific standards.

The next major acquisition involved not minerals, but rocks—of the extraterrestrial kind. In 1883, Harvard purchased a major meteorite collection from J. Lawrence Smith. Smith, a chemist and mineralogist from South Carolina, theorized that meteorites came from volcanic eruptions on the moon. He was mistaken, for later scientists established that nearly all meteorites come from asteroids. By the end of his life, however, Smith had built one of the finest collections of meteoritic stones in America, with specimens from some 250 falls.

In 1891, the century-long history of the Mineralogy Cabinet came to an end, and the collection officially became the Mineralogical Museum. The minerals (and meteorites) were moved to a new wing of the Oxford Street building that already housed the plant and animal collections. In their new gallery, the mineral specimens were installed in exhibit cases and systematically arranged, according to the fifth edition of James Dwight Dana's *Manual of Mineralogy*.

Some two decades later, Albert Fairchild Holden, an immensely successful mining engineer, raised Harvard's mineralogy collections to a new

Thomas Barbour with unidentified Choco Indians in Panama, 1922.

level of importance. Holden, managing director of United States Mining Company in Utah, assembled a large mineral collection over many years that he decided to leave to his alma mater. In 1910, knowing he was dying of stomach cancer, he aggressively purchased rare examples, intending to leave the finest possible legacy. The twenty-five hundred specimens he bequeathed in 1913 formed the basis of Harvard's modern mineralogical research collection. Holden's bronze memorial plaque in the Mineralogical Museum describes him this way : "Mining Engineer—Lover of Minerals—Administrator—Powerful and Eager in Many Pursuits." His name also lives on in the rare mineral that was named for him, holdenite.

Another major mineralogical benefactor was Albert Cameron Burrage, a wealthy lawyer and entrepreneur who graduated from Harvard in 1883. Burrage made his first fortune in the 1890s, representing the Brookline Gas Light Company during the "gas war" over Boston utilities. When his client won the case, Burrage received some $700,000 in fees. He then turned to mining, organizing both the Amalgamated Copper Company and the Chile Copper Company. He built a Boston Back Bay mansion, still standing on Commonwealth Avenue, bearing nearly two hundred griffins, dragons, gargoyles, and cherubs. In its billiards room, Burrage kept his enormous mineral collection, which included many rare and beautiful examples of crystallized gold. In 1947,

Burrage's wife, Alice Haskell Burrage, bequeathed to Harvard his collection, consisting of some five thousand specimens.

These benefactors helped make Harvard's mineral collection outstanding. The collection ranks among the top ten in the world, due to its breadth, the large number of specimens described in scientific literature, and its many rare minerals. The Hall of Minerals today displays to the public more than five thousand superb specimens, ranging from a stunning spiky red-orange crocoite from Tasmania to giant gypsum crystals from Mexico, standing more than five feet tall.

THE FROG TOSS. Alexander Agassiz died in 1910, ending the family's half-century-long leadership of the Museum of Comparative Zoology. From his death until 1927, the museum entered its dark ages—literally. The subsequent director, an aging entomologist named Samuel Henshaw, failed to modernize the building, which languished without electric lights. The exhibitions gathered dust.

Harold Coolidge with the gorilla he collected in eastern Congo, 1927.

Thomas Barbour, the museum's curator of herpetology and an extraordinarily energetic man, came to the rescue. He had already decided, at age thirteen, that he would grow up to become director of the Museum of Comparative Zoology.[38] Accepting the position in 1927, he immediately had electricity installed, revitalized the exhibits, and purged worthless specimens by placing them on the museum lawn, so that visitors could take them home for free. A self-described jack-of-all-trades, Barbour described new species of mammals, birds, reptiles, fish, and amphibians. He spoke fluent Spanish and made frequent scientific expeditions to Panama and to Cuba, where he helped develop Harvard's Atkins Institute, a botanical research station near Cienfuegos. (The institute was nationalized in the 1960s, after the Cuban Revolution.) Barbour wrote several wonderful books for popular audiences about his career, including *Naturalist at Large*, a pun on his considerable height and girth. He was, by all accounts, larger-than-life.

One of the best stories about Barbour is the frog toss. During the late 1930s, Barbour had a running debate with Harvard entomologist Philip Darlington over the origins of animal life in the West Indies. Barbour insisted that animals had arrived via land bridges, for he believed that the islands were once connected to one another and to Central America. Darlington disagreed, arguing that the animals could only have arrived by sea (rafting), or by air (carried by hurricane winds). Barbour contended that no animal could possibly survive a fall from a hurricane's heights. As a test, he proposed dropping frogs from the top of the five-story Museum of Comparative Zoology, certain that they would die.

Darlington climbed the many stairs, carrying several live frogs, and prepared to toss them out the window. Barbour stood below with a crowd of spectators. As each frog plopped down into the grass by Barbour's feet, Barbour shouted triumphantly up to Darlington at the window, "That one's dead!" But within minutes, the stunned frogs recovered and began hopping around. It was a victory for Darlington.

In the end, both Barbour and Darlington were right. Scientists have documented that, to cite one example, in 1995 more than a dozen green iguanas arrived on the island of Anguilla via a raft of hurricane debris. This would have pleased Darlington. But geologists have also shown conclusive evidence of land bridges. Barbour, too, would have been vindicated.

THE GREAT WHITE CONSERVATIONIST. Expeditions by Harvard scientists led to huge growth in its natural history collections, mainly from the mid-1800s to the start of World War II. (Although scientists continue to make collecting expeditions today, these tend to be more limited in duration and focused in scope.) Notable trips included Louis Agassiz's fifteen-month-long Thayer Expedition to Brazil in 1865–66, Alexander Agassiz's voyages on the *Blake* and *Albatross,* and Ernest H. Wilson's plant-collecting expeditions in the early 1900s to China.

In 1926, zoologist Harold J. Coolidge Jr., in the tradition of the great white hunter, set off on the Harvard African Expedition to shoot big game. In eastern Congo, his goal was to collect a mountain gorilla for the Museum of Comparative Zoology's collections. Coolidge climbed the steep, forested mountain ridges, waiting to get his shot. He twice wounded one of the big male gorillas, then finally got him in the heart. He wrote:

I had to take off my hat to this old king who handled his troops so well . . . and took his medicine like a man. The first bullet was in his shoulder, the second in his stomach, and the third had broken the jaw on the way to the heart. At this juncture the skies opened in

sympathy for him, and it has seldom, if ever, rained so hard.[39]

The gorilla, 5 feet 8 ¼ inches tall, now stands in a glass case in the Harvard Museum of Natural History. Coolidge estimated its weight at 475 pounds.

In 1928, Coolidge left on an expedition with Theodore Roosevelt Jr., son of the president, to explore Indochina. Then, in 1937, Coolidge led Harvard's Asiatic Primate Expedition (APE) into Thailand to collect gibbons, shooting whole family groups for study. Interestingly, like many big-game hunters of the era, Coolidge ultimately recognized the need for animal conservation. Coolidge went on to become a founder of the International Union for Conservation of Nature and Natural Resources, which publishes the "Red List" of endangered species, and in 1961 helped establish the World Wildlife Fund.

THE CONTRIBUTIONS OF WOMEN. Historically, women contributed to natural history endeavors at Harvard in various ways, working as assistants in collections, as writers, as illustrators, and as scientists in their own right. Generally, however, they did not hold doctorates and so were barred from professorships and most senior positions. As late as the 1960s, a woman might have met resistance when seeking to publish a scientific paper under her own name. Not until 1976 did a woman receive tenure at Harvard in the natural sciences. This was Ruth Dixon Turner, a Ph.D. and expert on the mollusks known as shipworms.

In the nineteenth century, Elizabeth Cabot Cary

Agassiz made her mark as an educator and writer. *The Atlantic Monthly* published several of her reports on natural history expeditions undertaken with her husband, Louis Agassiz. Her articles described such far-flung destinations as Amazonia, the Galápagos Islands, and a glacier in the Strait of Magellan, at the southern tip of South America. In this delightful passage, she reported on the Hassler Expedition to South America in 1872:

> We had dipped our dredges in various ocean depths from the West Indies to the southernmost limits of the continent; we had examined the moraines of ancient glaciers and the craters of extinct volcanoes on Patagonian shores, and hunted guanacos and ostriches on the adjoining plains; we had roused the penguins and cormorants by hundreds in their breeding-places on the cliffs of Magdalena Island, and seen the sea-lions lying on the beaches below, and so through manifold adventures by flood and field had come at last on a fine day in March to be lying off Glacier Bay in the Straits of Magellan.[40]

In 1865, Elizabeth Cary Agassiz helped Louis Agassiz organize the Thayer Expedition to Brazil, which yielded thousands of fish and other specimens for Harvard's collections. Together, they coauthored a popular book about the trip, *A Journey in Brazil.* Following her husband's death in 1873, she helped to found the Annex, the first women's school linked to Harvard. In 1894, this officially became Radcliffe College.

Another remarkable woman naturalist was Blanche Ames Ames, a gifted botanical illustrator.

In 1900, she married Harvard botanist Oakes Ames (not a relative) and collaborated with him for fifty years, as he became a world expert on orchids. Blanche Ames illustrated nearly all of Oakes Ames's scientific publications, including the seven-volume *Orchidaceae.* An 1899 graduate of Smith College, she was a suffragist, a cofounder of the Birth Control League of Massachusetts, and an inventor. During World War II, she received a patent for a method of stopping enemy airplanes by ensnaring their propellers in the wires of barrage balloons.

Elizabeth Bangs Bryant, born in 1875 and educated at Radcliffe, made her mark as an expert on West Indian spiders. She worked without pay for almost thirty years at the Museum of Comparative Zoology, before at last being granted a small stipend. Bryant published more than three dozen scientific papers, most written after the age of fifty-five. Her productivity was all the more remarkable given the inadequacy of her microscope, lit by a modified car headlight.[41] She stubbornly refused to be listed in American Men of Science, feeling that the honor belonged only to those professionally trained. At her death in 1953, she left a bequest to the Museum of Comparative Zoology for the care and study of its arachnids.

One other early woman scientist of note was Barbara Lawrence. After graduating from Vassar College in 1931, Lawrence became an assistant at the Museum of Comparative Zoology. She then embarked on an expedition to the Philippines and Sumatra to study bats. Two years later, she married William Schevill, then the museum's assistant curator of invertebrate paleontology. Together, in 1949, they made the first scientific recordings of porpoise sounds, using an underwater device called a hydrophone.[42] Lawrence served as curator of mammals at the Museum of Comparative Zoology from 1952 until her retirement in 1976. Her specialty became "zooarchaeology"—the study of animal remains from archaeological sites—and she published important articles on the origins of domestic dogs.

Decline and Rebirth

DNA AND THE DECLINE OF THE NATURALIST. In 1953, the discovery of the structure of DNA had troubling implications for biologists who studied "whole organisms." Many scientists saw the future as genetics, not taxonomy. The classifying of plants and animals—the very foundation of natural history museums—was increasingly considered a dead-end pursuit. Molecular biologists viewed scientists who studied the life histories of plants and animals as quaintly Victorian.

Edward O. Wilson wrote memorably about this period in his autobiography, *Naturalist*. Over his career, Wilson became one of the most articulate and effective champions of the importance of whole-animal studies. Here, he recalled his antipathy toward James Dewey Watson, who, with Francis Crick, had discovered the famous double helix, ushering in the molecular-biology revolution:

> When [Watson] was a young man, in the 1950s and 1960s, I found him the most unpleasant human being I had ever met. He came to Harvard as an assistant professor in 1956, also

my first year at the same rank. At twenty-eight, he was only a year older. He arrived with a conviction that biology must be transformed into a science directed at molecules and cells and rewritten in the language of physics and chemistry. What had gone before, "traditional" biology—*my* biology—was infested by stamp collectors who lacked the wit to transform their subject into a modern science. He treated most of the other twenty-four members of the Department of Biology with a revolutionary's fervent disrespect.[43]

Watson took an equally dim view of ecology. When Wilson suggested that Harvard add an environmental biologist to its faculty, Watson replied, "Anyone who would hire an ecologist is out of his mind."[44] (Over the years, Wilson and Watson became friends and even joined in public dialogues about the unity of biology, encompassing everything from molecules to ecosystems.)

In 1969, Harvard's Biology Department split in two, as the cell specialists formed a separate department for molecular biology. Then, in 1978, Biology split again. This time, the scientists who studied whole organisms created a separate Department of Organismic and Evolutionary Biology (OEB). It is the OEB scientists who have remained, over time, most closely associated with Harvard's natural history collections.

In the meantime, natural history collections at many universities across the country began to suffer from neglect. One symbol of their dire state was that in 1985, Princeton University simply gave away its entire collection of vertebrate fossils. Geologists there declared they were no longer interested in collections-based research, preferring to make room for high-tech laboratories. Princeton's large and historically significant collection was divvied up, with Yale's Peabody Museum alone receiving some fifteen thousand specimens. And Princeton was not alone in its efforts to downsize its natural history collections.

JURASSIC PARK AND THE REBIRTH OF NATURAL HISTORY. Then the unexpected happened. If science typically leaves old disciplines in the dust, in this case an older way of looking at the world was revalidated.

Starting in the 1980s, some biologists began to realize the limitations of molecular studies. No matter how much frog DNA they sequenced, that data alone could not explain why amphibians were suddenly dying off around the world. They needed traditional information about frog life cycles, species distributions, and anatomy. At the same time, scientists and others began expressing their concern about global habitat loss and a new mass extinction. In 1986, Edward O. Wilson introduced the term *biodiversity*—a portmanteau word catchier than *biological diversity.* His book *The Diversity of Life*, published in 1992, helped inspire a new generation of biologists with a mission: saving plants and animals in the face of human overpopulation and habitat destruction. In the 1990s, natural history museums also embraced conservation as a major theme, seeking to awaken the same kind of awareness in the broader public.

Meanwhile, in 1991 Michael Crichton published his blockbuster book *Jurassic Park*, popularizing the idea that dinosaurs could be resurrected using DNA preserved in amber. Although scientists gave little credence to Crichton's notion, they were in fact beginning to find DNA in a nonliving source: museum specimens. The first successful extraction of DNA from a dead specimen likely dates from 1980, when Chinese scientists sampled human remains nearly two thousand years old.[45] Then in 1985, biologists extracted DNA from herbarium specimens—dried, pressed plants.[46] Three years later, genetic material was successfully extracted from museum bird skins. With these advances, scientists began to see natural history museums in a new light, as vast libraries of DNA.

The DNA extracted from museum specimens has powerful applications. Biologists are using the data for a vast and ambitious project, building a great "tree of life." This tree, constructed over many years, aims to show each species' place in the grand scheme of evolution. Museum specimen DNA also helps scientists eliminate confusion, by providing information used to distinguish among species that appear similar to the naked eye. Additionally, the DNA from museum specimens is useful in conservation work. One researcher has, for example, sampled Galápagos tortoise shells collected in the nineteenth century to determine which species were native to particular islands. Based on this data, the tortoises—whose species have been muddled in captivity—could possibly be restored to their original habitats.

At Harvard, the fields of evolutionary biology and earth science were revitalized during the 1990s, attracting top-notch graduate students. Some three hundred researchers now work on a wide range of topics. One group studies bioluminescence, the ability of certain animals to produce light. Another looks at global change, particularly the effects of rising carbon dioxide levels. Certain researchers focus on astrobiology—the study of the origin, evolution, and destiny of life in the universe—while others look at how the earth's life-forms recover following mass extinctions. There are scientists investigating the mechanics of how wallabies hop, how fish swim, and how movement evolved in mammals over millions of years. Others study such topics as carnivorous fungi, plants' responses to environmental stress, and the transformation of New England forests over time. Typically, Harvard scientists use a mix of techniques, combining molecular biology with more traditional approaches. Evolutionary investigation underlies many research projects, as scientists seek to understand where species came from and where they are headed.

THE CONVERGENCE: WHERE OLD AND NEW BIOLOGIES MEET. Roger Vila Ujaldón is typical of a new generation of researchers around the world, comfortable both in the field and the laboratory. A native of Barcelona, he works as a postdoctoral fellow in the lab of Naomi Pierce, a distinguished Harvard butterfly expert who combines evolutionary and molecular biology in her work.

"I study butterflies," Vila tells people who ask about his job. This is a useful conversation opener, for almost everyone has a story to tell about butterflies. If he starts out by saying, "I study molecular evolution," the conversation generally stops right there. What Vila does—working closely with others in the lab—is in fact quite complex,

combining traditional biology with butterfly genetics. He is, however, skilled at explaining his work in simple terms. By analyzing butterfly DNA, Vila may succeed in distinguishing a new endangered species that deserves protection. And by constructing charts called phylogenetic trees, he may come to understand how groups of butterflies spread and evolved over millions of years.

Inside tiny vials, Vila has preserved butterfly bodies in alcohol. From these specimens, he can analyze the genetic sequences within butterfly cells, using a technique called the polymerase chain reaction (PCR). PCR was perfected in the mid-1980s. The technique revolutionized the field of biology by making it possible and affordable to duplicate segments of DNA. Following this breakthrough, many scientists who had always studied plants and animals in traditional ways began integrating molecular biology into their research.

When Vila compares the DNA sequences of butterflies, he sometimes gets surprising results. A butterfly that has long been lumped together with a similar species may prove to have quite distinct DNA. It may, in fact, deserve to be considered a species in its own right. If the butterfly is quite rare, it might also earn official listing as an endangered species, providing protection that it would not otherwise have received.

Vila and other zoologists around the world also use another established technique: they construct phylogenetic trees. These are like people's family trees, for they reveal ancestry and degrees of relatedness among kin. The trees are based on DNA sequences. They, too, can be powerful tools. Phylogenetic trees can reveal much about where

Fossil skeleton being moved to an upstairs gallery, 1959.

animals first appeared, how they spread, and how they evolved over millions of years.

Vila has had a somewhat unusual scientific career, but his motivations are shared by many biologists. "Usually," he explained, "people who had been studying or doing research in classical biology end up discovering the potential of new molecular tools. My case is the contrary. I studied biochemistry and molecular biology, and my Ph.D. was focused on protein structure. However, since childhood I have always been mad about butterflies, and there was a moment in my life, just a few years ago, when I realized that my skills could perfectly match my passion for butterflies! Before, I always said that I should have been born in Darwin's age, when an entomologist had everything to discover. But maybe I was born at just the right moment in the history of science, when molecular and traditional studies are converging, leading to a better understanding of nature and ourselves."[47]

A New Public Focus

THE HARVARD MUSEUM OF NATURAL HISTORY. For decades, the public displays of Harvard's natural history collections languished. The Glass Flowers grew dusty and the stuffed animals shabby. The science depicted was often years out-of-date. Each of the three separate museums—specializing in animals, minerals, and plants—had only a modest staff for public programming and outreach. There was no capacity to undertake major exhibitions or badly needed conservation work. Limited effort was made to draw in visitors, and outside of Cambridge and scientific circles, the museums were something of a well-kept secret.

Then in the early 1990s, a group of scientists and administrators decided to take action, aiming to make the public mission a priority. They sought to cut across intellectual and organizational boundaries, to better present Harvard's collections and scientific research to the outside world. This initiative led, in 1995, to the formation of the Museum of Cultural and Natural History, which included the participation of the Peabody Museum of Archaeology and Ethnology. The new public institution was renamed the Harvard Museum of Natural History in 1998, when the Peabody withdrew from the collaboration.

The Harvard Museum of Natural History brings together the public sides of the three original collections—the Museum of Comparative Zoology, the Harvard University Herbaria, and the Mineralogical Museum. Since its creation, every aspect of public programming has been revitalized. The museum has created lecture series, renovated galleries, and expanded educational outreach. Exhibitions highlighting the work of Harvard scientists—on topics ranging from the sex lives of orchids to the origins of life on earth—have brought in new visitors. The museum has undertaken a variety of initiatives aimed at educating citizens about the current biodiversity crisis and the great conservation challenges lying ahead.

The Harvard Museum of Natural History draws on collections containing more than 21 million specimens, constituting one of the largest and most important natural history repositories in the world. Its mission is to "enhance public understanding and appreciation of the natural world and the human place in it, sparking curiosity and a spirit of discovery in people of all ages."[48] *The Rarest of the Rare* is part of that endeavor.

How to Read This Book

Behind every specimen in this book is a good story. There are tales of wealthy explorers, obsessive collectors, bone hunters, mushroom seekers, and visionary scientists. The specimens themselves are immensely varied and appealing. They come from the farthest reaches of the globe, the deepest depths of the sea, and even outer space. Some are beautiful (the tanagers, the gold). Others are intriguing (the dodo skeleton, the fossil of the gigantic dragonfly wing). Others are heartbreaking (the wolf pelt, the extinct butterfly). And still others are simply strange (the tapeworms from the digestive tracts of upper-crust Bostonians, the birdwing butterfly collected by a man later eaten by cannibals).

The following six chapters are organized by theme, rather than by taxonomic grouping, to provide an evocative tour of the collections. Each photograph is accompanied by a short essay. Captions indicate scientific names and collection dates, if known. The specimens' museum catalog numbers can be found in the endnotes.

The Harvard Museum of Natural History is a place of science, and yet it also conveys—for anyone drawn to adventure and discovery—an undeniable romance. This book seeks to capture something of both.

1 Meriwether Lewis's Last Bird
& Other Historic Holdings

. . . by day and by night, as long as I could keep my eyes open, did I labour to preserve animals, common as well as those more rarely met . . .

—Charles Willson Peale[1]

Peale, who founded America's first scientific museum in Philadelphia in 1784, taught himself the art of taxidermy. His early attempts were not entirely successful, and he once had to discard an Angora cat donated by Benjamin Franklin. But he soon developed his own taxidermy techniques, using arsenic solutions, that he considered superior to any known in Europe. When George Washington sent him a pair of golden pheasants in 1787, Peale preserved the birds so admirably that they can still be exhibited at the Harvard Museum of Natural History, more than two hundred years later.

Harvard has many remarkable historic holdings, some of them showcased in this chapter. One particular favorite is the mamo, a Hawaiian bird collected in the 1770s during Captain James Cook's third voyage of discovery. Another is the Lewis's woodpecker collected during the Lewis and Clark expedition, believed to be the only complete animal specimen remaining from that landmark exploration of the American West.

Indeed, the bird collection at the Museum of Comparative Zoology contains so many historic specimens that making a selection proved difficult. Harvard has, for example, dozens of specimens that once belonged to Peale's Museum, which survived until about 1849. From the Peale collections came the Lewis's woodpecker, as well as many important bird specimens studied in the early 1800s by Alexander Wilson, the father of American ornithology.

Birds are not all, however. Henry David Thoreau's herbarium—his lifetime collection of dried, pressed plants—is at Harvard, a notable legacy of this brilliant writer and ecologist. In addition, Harvard has turtle embryos studied by Louis Agassiz, the visionary founder of the Museum of Comparative Zoology, as well as deep-sea invertebrates collected by his son, Alexander Agassiz, who took over leadership of the museum and became a respected oceanographer in his own right.

There is a place in natural history collections for both old and new. People often ask why natural history museums collect so many examples of the same species, and why they continue to add to their collections over time. The answer lies in the

A Blanding's turtle collected by Henry David Thoreau. It was almost certainly among the fish and turtle specimens from Walden Pond that he sent to Louis Agassiz at Harvard in June of 1847.

nature of scientific research. To fully understand evolution and other biological processes, scientists need to be able to compare specimens from different locations, different conditions, and different eras. A Lewis's woodpecker collected today has all the more value if it can be compared to the specimen brought back from the Lewis and Clark expedition, displayed long ago in the Philadelphia museum of Charles Willson Peale.

Captain Cook's Mamo

Drepanis pacifica (dre-PAN-is), collected about 1778.[2]

> **For who, having a spark of imagination, could fail to be thrilled to hold in his hand our specimen of *Drepanis pacifica*?**
>
> —Thomas Barbour, director of the Museum of Comparative Zoology, 1943[3]

This mamo was collected in Hawaii during Captain James Cook's third and final voyage of discovery. The English explorer and his crew landed on the island of Hawaii in 1778, likely becoming the first white visitors. At first, Cook was welcomed by the Hawaiian people, but relations soon soured. In February of 1779, he was forced to return to the island after a violent storm damaged his ship. During fighting between Hawaiians and his crew, Cook was stabbed to death.

Mamos, now extinct, once inhabited the Hawaiian forests, where they used their long, curved bills to sip nectar from the abundant flowers. Their few yellow feathers made them valuable to the Hawaiians, who used them as symbols of nobility. In the early 1800s, King Kamehameha I wore a yellow cloak made from the feathers of some eighty thousand mamos. Professional bird catchers would capture the mamos, pluck the yellow tufts from their bodies, then set them free. It is unclear whether the birds survived the plucking. Scientists generally blame the mamo's extinction on introduced diseases and habitat loss, not on feather collecting. The last recorded sighting of a mamo was in 1898.

This specimen's provenance is unusually interesting. When Cook's ship returned home, the mamo was given to the Leverian Museum in London. The Leverian, a private museum owned by Sir Ashton Lever, comprised sixteen rooms filled with bird specimens, fossils, and, for intrepid visitors, a room of "monkies and monsters." Among the museum's treasures were a wealth of Pacific specimens from Captain Cook's earlier voyages of discovery.

In 1786, Sir Ashton disposed of his fabulous collection—including the mamo—by lottery. Twenty years later, the new owner auctioned off the entire contents of the museum, during a sale that lasted sixty-five days. Among the bidders was the naturalist Leopold von Fichtel, commissioned to buy specimens for Emperor Francis I of Austria. Fichtel purchased this mamo and a second one, which entered the holdings of Vienna's Naturhistorisches Museum.

Years later, the mamo was obtained from Vienna by an American bird collector, Leonard C. Sanford. He gave it to Harvard in exchange for a rare specimen of *Ciridops*, another Hawaiian honeycreeper that became extinct.

Minerals out of History

Serpentine, donated by J. C. Lettsom in 1793.[4]

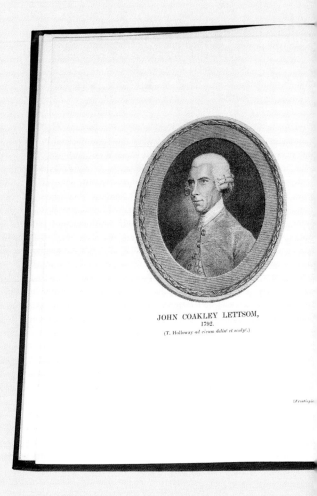

JOHN COAKLEY LETTSOM,
1792.
(T. Holloway ad vivum delin' et sculp'.)

[Frontispie.

Harvard's first scientific collection, the Mineralogical Cabinet, owes its existence to Benjamin Waterhouse. Waterhouse, a brilliant but cantankerous physician, is mainly remembered as the man who introduced the smallpox vaccine to the United States. Yet he was also a significant early mineralogist, with important ties to Europe.

In 1775, with war looming, Waterhouse left for England on the last ship out of Newport, Rhode Island. In London, he struck up a friendship with a fellow Quaker, John Coakley Lettsom. Lettsom, a prominent English physician and philanthropist, had American sympathies. The two corresponded regularly after Waterhouse returned to the independent United States, becoming a professor at the newly founded Harvard Medical School.

In 1793, Lettsom made a generous gift to Harvard: seven hundred mineral specimens from all across Europe. This was one of the first systematic collections of minerals in the United States. Among them was the green serpentine shown here, bearing its original label in Waterhouse's handwriting. The serpentine, a decorative stone, came from Banffshire, Scotland.

Waterhouse arranged for Lettsom's minerals to be displayed in Harvard Hall, in an eighteen-foot-long mahogany cabinet. George Washington came to the see them, as did Daniel Webster. But regular farmers also paid visits, traveling long miles to show Waterhouse their "gold." "I undeceived these people in as gentle a manner as I could," Waterhouse wrote to Lettsom. "Then I showed them true gold and silver ores from your collection, and contrasted them with the worthless but glittering pyrites . . ."

LETTSOM

HIS LIFE, TIMES, FRIENDS
AND DESCENDANTS

LONDON
WILLIAM HEINEMANN
MEDICAL BOOKS LTD
1933

Barites, given in the 1790s, with antique French
Revolution postcard.[5]

Waterhouse was by all accounts a prickly and
pedantic man. He dressed in the style of an old
English physician, complete with powdered
pigtail and gold-topped cane. He left Harvard in
1812, shortly after losing his professorship at
the medical school due to feuds with colleagues.
He apparently took the earliest catalogs of the

Mineralogical Cabinet with him, as they have been
missing ever since.

The two barite specimens shown here are also
historic. In 1795, the new French Republic presented Harvard with a gift of 189 mineral specimens, including these barite crystals from Saxony,
Germany. The accompanying letter brims with
revolutionary fervor and fraternal feeling.
Addressed to the "Citizen brothers" of Harvard,
it begins: "your Example has inflamed our Souls
with the Sacred fire of Liberty."

Meriwether Lewis's Woodpecker

Melanerpes lewis (mel-an-ER-peez), collected in 1806.[6]

In 1803, President Thomas Jefferson charged Meriwether Lewis with the mission of finding a river route to the Pacific Ocean. Along the way, Lewis was to collect animal, mineral, and plant specimens from the unexplored West. Jefferson, who resisted the idea of extinction, thought Lewis might well find woolly mammoths living on the land.

Although mammoths proved elusive, Lewis did procure animals new to science, including a coyote and a prairie dog. He also obtained novel species of birds. On May 27, 1806, camped on the Clearwater River in what is now Idaho, Lewis noted in his journal (his spellings preserved):

> The Black woodpecker which I have frequently mentioned and which is found in most parts of the roky Mountains as well as the Western and S.W. Mountains, I had never an opportunity of examining untill a few days since when we killed and preserved several of them. . . . The belly and breast is a curious mixture of white and blood red which has much the appearance of having been artificially painted or stained of that colour.

The bird now bears his name: Lewis's woodpecker, or *Melanerpes lewis*. Of the many animals collected during the Lewis and Clark expedition, this is believed to be the only complete specimen remaining. (Also surviving are elk antlers, displayed at Monticello, and the horn of a bighorn sheep, held by the Filson Museum in Louisville, Kentucky.)

In all likelihood, Lewis carried this very woodpecker back for Charles Willson Peale to display in his extraordinary museum of art and natural science in Philadelphia.[7] Peale's Museum, founded in 1784, served as a national repository before the creation of the Smithsonian Institution. It was at Peale's Museum that Alexander Wilson saw this specimen. He used it as the model for his illustration of Lewis's woodpecker in *American Ornithology,* his landmark nine-volume study of North American birds.

In about 1849, the contents of the Peale museum were sold to showmen Moses Kimball and P. T. Barnum. Kimball's share—with the woodpecker—went to the Boston Museum, and then to the Boston Society of Natural History. While cleaning house, that society sold off the dusty bird specimens to one C. J. Maynard, who stored them in his barn. A few years later, the museum reclaimed the birds and finally transferred them, in 1914, to Harvard's Museum of Comparative Zoology.

Cyclopterus lumpus [him]

Collection Prof. Peck. 1793
Cyclopterus Lumpus
Lump Sucker

Fish Prepared the Linnaean Way

Cyclopterus lumpus (sy-KLOP-ter-us), collected in 1793.[8]

This strange dried fish—a lumpfish—has tremendous historical value. It was collected in 1793 by the American naturalist William Dandridge Peck. Harvard still has more than a dozen fish that Peck preserved this way more than two centuries ago, comprising one of the earliest biological collections in the United States.

Peck preserved the fish according to a Linnaean method: they were skinned, sliced in half lengthwise, dried and pressed, then mounted on paper. Linnaeus had trained as a botanist, and this method treats fish almost as if they were plants. At the Linnean Society of London, there exist similar dried and pressed fish specimens from Linnaeus's personal collection, also dating back to the 1700s. Today, fish specimens are nearly always preserved in alcohol.

Peck's career can be traced to Linnaeus, for it is said that he first grew interested in natural history after a copy of Linnaeus's *Systema Naturae* washed up near his house following a shipwreck. Peck went on to study classics and natural history at Harvard, graduating in 1782. In the 1790s, he first made his mark as a scientist with his "Natural History of the Slug Worm."

Not to be overlooked is his paper "Description of four remarkable Fishes, taken near the Piscataqua in New Hampshire." This was among the earliest papers published in the United States on systematic zoology, the science of classifying animals. Peck's article was published in 1804 in the distinguished journal *Memoirs of the American Academy of Arts and Sciences.*

The following year, Peck was appointed Massachusetts Professor of Natural History at Harvard. During his career, he focused mainly on entomology, publishing on insects that attacked pear, oak, cherry, and locust trees. Peck sent various insect specimens and descriptions to William Kirby, an English entomologist, who published some of them in the *Transactions* of the Linnean Society of London. Kirby named an insect that lives in the abdomen of wasps *Xenos peckii,* in his honor.

It is amusing to note that, for all his erudition, Peck's second contribution to the *Memoirs of the American Academy* was titled "Some Observations on the Sea Serpent." The paper, from 1818, presents accounts by "men of fair and unblemished character" of a sea serpent at least seventy feet long, fast-swimming and nearly black in color, seen off the coast of Massachusetts.

Asa Gray's Long-Lost Flower

Shortia galacifolia (SHORT-ee-ah), collected in 1787.[9]

Asa Gray, the foremost American botanist of the nineteenth century, sought this plant for years but could not find it. The tale of the lost *Shortia* is long and captivating. It began in 1787, when French botanist André Michaux collected an unknown plant while exploring the southern Appalachians. He pressed and dried the specimens, and they became part of his herbarium in Paris.

Some fifty years later, in 1839, Gray came across the unnamed specimen in Michaux's collection while doing research in Europe. He made the note, shown here, that Michaux had found the flower in the "Hautes Montagnes de Carolinie," the high Carolina mountains. Following this vague indication, Gray was determined to find the plant growing in the wild.

He searched the Carolina mountains without success. Three years later, Gray decided to name the plant for science anyway, relying on one of Michaux's dried specimens, pictured here. He called it *Shortia*, after Kentucky botanist Charles W. Short.

In 1858, the *Shortia* story took another turn. Gray, who had joined the Harvard faculty in 1842, came across a Japanese plant nearly identical to the American *Shortia*. He observed that a surprising number of plants were found in only two places on earth: East Asia and the American Southeast.

These were, he believed, relics of a preglacial flora that was once widespread.

Gray's theory renewed interest in finding Michaux's original *Shortia*. Botanists combed the Carolina mountains in vain. Finally, in 1877, a seventeen-year-old named George Hyams discovered a likely patch growing near Marion, North Carolina. When flowering specimens were collected the following spring, Gray confirmed their identity.

Gray, by this time, had become one of the best-known American scientists, and one of the first to establish a reputation in Europe. He was famous, too, as an outspoken defender of Darwin's theories. Among his legacies is Harvard's Gray Herbarium, established with his personal collection of dried, pressed plants.

The final chapter in the *Shortia* story unfolded in 1886, near the end of Gray's life. Charles Sprague Sargent, a student of Gray's and the first director of Harvard's Arnold Arboretum, set out to find the plant, using Michaux's journal as his guide. Sargent found the plant growing near Highlands, North Carolina—a place, incidentally, that the glaciers had never reached. At last, nearly a century after Michaux's discovery, the lost *Shortia* was found.

s Carolinæ Septentriona
egit A. GRAY et J. Ca

Plante (sauter? Mr Michx

"Hautes Montagnes de Caroline"
Caroline." Michx —
"An Pupole spec. an genus novum."

Thoreau's Wild Cranberry

Vaccinium oxycoccus (vac-SIN-ee-um), collected after 1850.[10]

Henry David Thoreau wrote passionately about cranberries. He found this kind, the "small cranberry," growing wild in his hometown of Concord, Massachusetts. The sheet shown here is from his personal herbarium, his collection of dried, pressed plants. In pencil, he has noted the cranberry's location—Gowing's Swamp—and its scientific name, *Vaccinium oxycoccus.* (Thoreau's handwriting is notoriously difficult to read.)

In the last years of his life, Thoreau, who was a Harvard graduate, became increasingly serious about botanizing. He left an unfinished manuscript that focused exclusively on plants from the Concord area. Published in 2000 as *Wild Fruits*, the work displays Thoreau's usual grace and spirit. The section on the small cranberry is quite memorable.

On October 17, 1859, Thoreau spent the afternoon "a-cranberrying," aiming to have "a dish of this sauce on the table at Thanksgiving of my own gathering." He noted that although this effort earned him not a penny, he received a different kind of recompense. His cranberries conveyed the very essence of Gowing's Swamp "and of *life* in New England." He wrote:

> Let not your life be wholly without an object, though it be only to ascertain the flavor of a cranberry, for it will not be only the quality of an insignificant berry that you will have tasted, but the flavor of your life to that extent, and it will be such a sauce as no wealth can buy.[11]

Thoreau's life was brief, for he died of tuberculosis at age forty-four. Just two of his major books were published while he was alive, *A Week on the Concord and Merrimack Rivers* and *Walden.* His real fame was posthumous; a man a century ahead of his time, he is often considered the first ecologist.

Wilson's Pet Parakeet

Alexander Wilson may lack the name recognition of John James Audubon, but bird experts everywhere know Wilson as "the father of American ornithology." Born in Scotland, Wilson began his career as a weaver and poet, then emigrated in 1794 to the United States. He settled in Philadelphia, and there he wrote his monumental nine-volume *American Ornithology*, the first comprehensive study of North American birds.

In producing his books, Wilson acted as researcher, writer, and illustrator. He traveled thousands of miles, often on foot, collecting specimens and observing birds in the field. Pictured here is his lovely hand-colored plate of the Carolina parakeet from volume 3 of his opus, published in 1811. With it is the very specimen that Wilson used as a model for his illustration.

Wilson had a particular fondness for Carolina parakeets. In *American Ornithology*, he described seeing a great flock of them at Big Bone Lick in northern Kentucky:

> They came screaming through the woods in the morning, about an hour after sunrise, to drink the salt water, of which they, as well as the Pigeons, are remarkably fond. When they alighted on the ground, it appeared at a distance as if covered with a carpet of the richest green, orange, and yellow . . .[13]

Wilson even kept one Carolina parakeet—though not the individual shown here—as a pet. The bird, which he had slightly wounded in the wing, became his "sole companion in many a lonesome day's march." While traveling on horseback, Wilson bound the parakeet in a silk handkerchief and secured it in his pocket, liberating it at mealtimes and in the evening. "When at night I encamped in the woods," he wrote, "I placed it on the baggage beside me, where it usually sat, with great composure, dozing and gazing at the fire till morning."

Wilson would have been saddened to learn that the Carolina parakeet has become extinct. The last one died in its cage at the Cincinnati Zoo in 1918. The flocks that once ranged over much of the eastern United States disappeared as forests gave way to cropland. Farmers shot the birds to protect their orchards, hunters shot them for sport, and their feathers were used in hats. They were the only parrot native to eastern North America.

Drawn from nature by A. Wilson   Engraved by I. Seton

1. *Carolina Parrot.* 2. *Canada Flycatcher.* 3. *Hooded F.* 4. *Green, black capt. F.*

Louis Agassiz's Turtle Embryo

Embryo of *Chelydra serpentina* (KEL-i-dra), dated 1852; plate from *Contributions to the Natural History of the United States of America*.[14]

In 1848, Louis Agassiz left his native Switzerland for good to become a professor at Harvard. Already, he was well-known for his brilliant work on glaciers and fossil fish.

In America, he would go on to many impressive accomplishments, founding the Museum of Comparative Zoology and training an entire generation of zoologists. If his greatest discoveries were behind him, he nonetheless produced one notable scientific work during his Harvard years, which spanned from 1848 until his death in 1873. This was *Contributions to the Natural History of the United States of America*. The first two volumes, published in 1857, comprised an exhaustive study of turtles, with fine illustrations by Jacques Burkhardt, Henry James Clark, and Auguste Sonrel.

Shown here is a historic turtle embryo, used as a model in Agassiz's book. With it is the corresponding illustration, depicting turtle embryos and eggs. The embryo is that of *Chelydra serpentina*, a common snapping turtle.

Before writing *Contributions*, Agassiz had announced his intention to review all the turtle species of North America. To this end, he encouraged natural history enthusiasts across the continent to send him specimens. Turtles arrived from Toledo, Toronto, and Texas. Henry David Thoreau sent a Blanding's turtle from Walden Pond.

In 1910, *The Atlantic Monthly* published a delightful article about Agassiz's undertaking titled "Turtle Eggs for Agassiz." It described a time in the 1850s when Agassiz's work had ground to a halt, for he could not find freshly laid eggs revealing the earliest stages of turtle development. At last, a school principal from the southeast Massachusetts town of Middleboro came to the rescue. Determined to bring Agassiz eggs less than three hours old, John Whipple Potter Jenks tracked a female turtle through the mud before dawn, grabbed her eggs, galloped off on his horse, and intercepted a freight train headed for Boston. When he arrived breathless and filthy at Agassiz's house in Cambridge, just before seven o'clock on a Sunday morning, the maid tried to send him away. But Agassiz recognized him, Jenks recalled, and commanded:

"'Let him in! Let him in. I know him. He has my turtle eggs!'"

"And the apparition, slipperless, and clad in anything but an academic gown, came sailing down the stairs.

"The maid fled. The great man, his arms extended, laid hold of me with both hands, and dragging me and my precious pail into his study, with a swift, clean stroke laid open one of the eggs, as the watch in my trembling hands ticked its way to seven—as if nothing unusual were happening to the history of the world."[15]

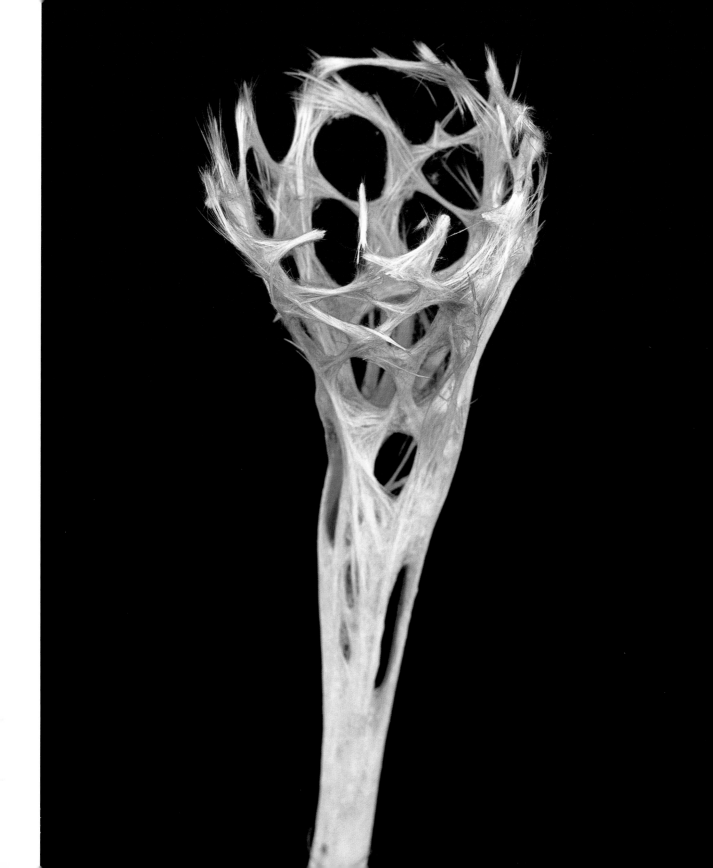

Alexander Agassiz's Glass Sponge

Rhabdopectella tintinnus (rab-doh-pec-TEL-a), collected in 1879.[16]

Alexander Agassiz, son of Louis, did not possess his father's enormous charisma. He did, however, make a name for himself as a successful businessman. In 1871, Alexander Agassiz became president of the Calumet and Hecla copper mine on Michigan's Upper Peninsula, an enterprise that soon turned immensely profitable. Over the years, he donated more than $1 million to the museum that his father had founded, Harvard's Museum of Comparative Zoology.

This is not to dismiss Alexander Agassiz's important contributions to science, particularly oceanography. In the nineteenth century, scientists investigating the deep sea faced considerable challenges. They had no submersibles, and no way of seeing what lay in the depths. Their only tool was dredging—lowering down a trawl bag, dragging it along the ocean bottom, and hauling it up to see what animals had been captured.

This painstaking process was further frustrated by the limitations of ropes, for even the best hemp ropes often snapped under the strain. Alexander Agassiz knew of a better alternative, from his experience in mining: steel rope. Collaborating with naval commander Charles Sigsbee, the two men created a dredging and anchoring system using steel cables. They put their invention to use in the winter of 1877, during a U.S. Coast Survey expedition on the steamer *Blake*.

The glass sponge pictured here was dredged using their innovation. Agassiz collected the specimen off Grenada during a subsequent voyage of the *Blake*, in 1879, at a depth of about seventeen hundred feet. He later published a classic two-volume work about the many discoveries from these expeditions, *Three Cruises of the Blake*.

Glass sponges bear little resemblance to familiar bath sponges. They are composed of a different material—silica, the very material used in making glass. This one, *Rhabdopectella tintinnus*, is a beautiful specimen, yet it is best handled wearing gloves. Biologists have little information about the species, which is known from only a few specimens.

Sponges—including glass sponges—are among the simplest animals on earth, having neither nerves nor muscles. Like plants, they remain rooted to one location. Certain glass sponges are, however, capable of responding to danger, as researchers at the University of Victoria in Canada have discovered. Biologists Sally Leys and George Mackie found that glass sponges can generate electrical impulses that "turn off" the sponge, shutting down its food-filtering system when the surrounding water becomes murky. This prevents sand and other large particles from clogging the sponge's pores.

The Secrets of Fish

Hoplostethus pacificus (hop-loh-STEE-thus), collected in 1891.[17]

Samuel Garman was one of the more eccentric characters at the Museum of Comparative Zoology. Although he ultimately specialized in fish, his first love was fossils. He participated in an early exploration of Colorado and, in 1872, served briefly as field assistant to dinosaur hunter Edward Drinker Cope. (Cope, in the famous "bone wars" with Othniel Marsh, was not above stealing his rival's research.)

In 1870, Garman read in a newspaper that Louis Agassiz would be landing at San Francisco at the end of the *Hassler* expedition to South America. On the wharf, Garman introduced himself, telling Agassiz about his fossil-collecting experience and his desire to be a scientist. Agassiz was so impressed that he invited Garman to Harvard, where he remained for the rest of his life. Garman took up deep-sea fish and sharks as his specialties. Perhaps his most important work was the beautifully illustrated 1899 book *The Fishes*.

Garman's book describes fish collected during voyages of the *Albatross*, a steamer owned by the U.S. Fish Commission. Although he did not go along on the expeditions, Garman received hundreds of fish specimens for study. He spent seven years working on his book and described a staggering 174 species of fish that were new to science, including the one pictured here, a deep-sea roughy. The roughy was collected on the *Albatross* in 1891, at a depth of 171 fathoms—about a thousand feet below sea level. Although alcohol has drained away its orange and yellow coloration, the illustration from *The Fishes* reveals the roughy's original beauty.

In his later years, Garman became rather secretive, perhaps because of his experiences with Cope. Thomas Barbour, director of the Museum of Comparative Zoology from 1927 to 1946, described Garman's strange habits with great bemusement in his book *Naturalist at Large:*

> He had quarters in the basement of the Museum which could not be reached except by a grilled door. One rang a bell, there was a rustle of papers (the shades were never pulled up, so you couldn't look in the windows); after a while Garman came to the door and opened it, the grille outside meantime being fastened. After he had verified the identity of his visitor, he might let him in and be quite friendly. Just as often he was too busy, and closed the door.[18]

That rustle of papers? According to Barbour, Garman was covering his worktable with newspapers, to prevent spying.

2 Charles Darwin's Buried Treasure
& Other Fossil Finds

Fossil collecting is hot, dirty, and discouraging work. But the results are worth the effort, and it's grand fun.

—Alfred S. Romer[1]

At first glance, the obvious fossils to include in this book appeared to be big dinosaurs with familiar names. The Museum of Comparative Zoology has, for example, the skeleton of a *Deinonychus* and the skull of a *Triceratops*. But several experts advised otherwise, saying these were not Harvard's most interesting specimens.

For decades, Harvard paleontologists have focused their research on the fascinating transition between reptiles and mammals. Accordingly, two mammal-like reptiles were selected for this chapter, *Kayentatherium* and the *Dimetrodon*. Neither is a dinosaur—for the word *dinosaur* refers to a specific group of ancient land-dwelling reptiles that excludes the mammal-like species. (Dinosaurs share such characteristics as upright posture, a vertical pelvis, and a hole in the hip socket.)

Other specimens were chosen for this chapter because they rarely appear in the fossil record, due to the softness or fragility of the animals' bodies. These include an ancient squid, a gigantic dragonfly, and an exquisite 35-million-year-old butterfly. As for the fossil sand dollar collected by Charles Darwin during the voyage of the *Beagle*, it speaks for itself.

Two fossils represented in the chapter are enormous—though neither is a dinosaur. One is a huge ancient turtle from South America, a *Stupendemys*, and the other a gigantic marine reptile collected in Australia, a *Kronosaurus*, which belongs to a group of predators called pliosaurs.

As for Harvard's *Triceratops*, it once had considerable importance as a type specimen, an individual chosen to represent a new species. Most experts now believe, however, that the species it represented, *Triceratops eurycephalus*, is actually a variant of a different species, *Triceratops horridus*. Such changes in taxonomy occur often in biology. In fact, taxonomists are themselves often classified into two groups, lumpers and splitters. Lumpers like to group variants together into the fewest possible species, while splitters prefer to define every variant as a distinct species. In 1986, Harvard's *Triceratops eurycephalus* fell victim to distinguished lumpers in Germany and at Yale University.

And so, there are no dinosaurs in this book. For readers who have always lumped all ancient vertebrates together as "dinosaurs," this chapter offers a modest lesson in splitting.

Darwin's Sand Dollar

Iheringiella patagoniensis (ee-rin-jee-EL-a), collected in 1834.[2]

Charles Darwin collected this fossil sand dollar in 1834, during the voyage of the *Beagle*. Hired as the ship's naturalist, he spent five years circumnavigating the globe. At every port, he collected natural history specimens—from birds to reptiles to fossilized mammal bones—and sent them back to England.

During the winter of 1834, the *Beagle* was descending the coast of what is now Argentina, stopping at various points in Patagonia. At Puerto San Julián, Darwin collected this fossil sand dollar. The ship docked there on January 9 and remained for eight days.

"The geology of Patagonia is interesting," Darwin wrote in *The Voyage of the Beagle*,[3] noting that the shores were lined with huge deposits of shells from extinct sea animals. This shell bed, overlain with white gravel, extended for hundreds of miles along the Atlantic coast. At Puerto San Julián, he observed, "Its thickness is more than 800 feet!" In his investigations of the shell bed, Darwin found one layer abounding with fossil sand dollars.

The history of this specimen is itself interesting. It appears likely that Darwin sent it to Louis Agassiz in Neuchâtel, Switzerland, where Agassiz taught natural history before coming to Harvard. Darwin sent many *Beagle* specimens to experts for identification, and Agassiz was well-known for his work on sand dollars and other echinoids.

The specimen was then named by Edward Desor, who had been Agassiz's right-hand man in Switzerland, but who became his bitterest enemy. In 1847, Desor brought the extinct sand dollar to a meeting of the French Société Géologique, where he gave it the name *Echinarachnius juliensis*. That name still appears on the old label pictured here.

That is not the end of the story, however. Later in 1847, the sand dollar was formally described in a publication called *Catalogue raisonné . . . des échinodermes*. Although that catalog's authors are listed as both Agassiz and Desor, Agassiz insisted that the work was his alone. Desor had merely claimed authorship "with a stroke of his pen—in my absence," Agassiz angrily wrote.[4] On this point, Agassiz was eventually vindicated.

In 1898, the sand dollar's name was changed to *Iheringiella patagoniensis*, after Hermann von Ihering, an expert on the fossil shells of Patagonia. The name change occurred because Desor had described two species that turned out to be "synonyms"—the same species. No doubt Agassiz would have been pleased that Desor's name did not survive.

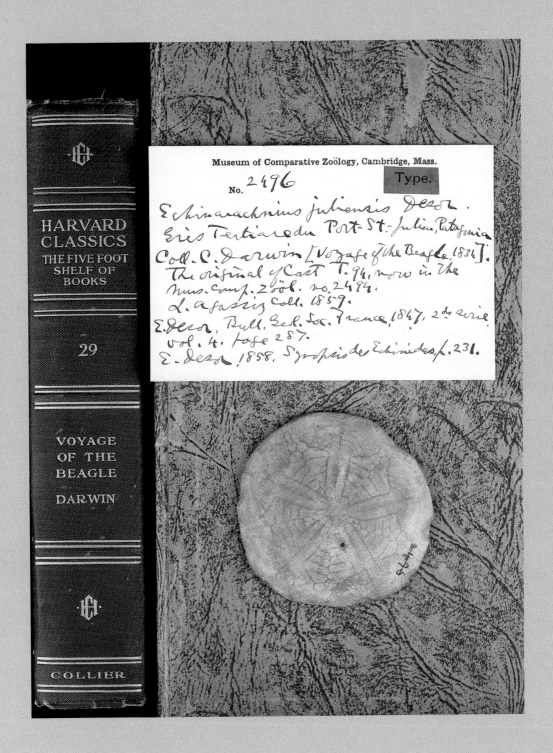

The Mastodon Murder

Composite skeleton of *Mammut americanum* (MAH-moot), collected in 1844.[5]

Idocrase (EYE-doh-craze), now called vesuvianite, likely collected by John White Webster in 1848.[6]

For John White Webster, this mastodon was too good to resist. He knew the huge skeleton would make a dramatic center-piece for Harvard's "cabinet"—its still modest natural history museum. The mastodon, a rare nearly complete specimen, was excavated in 1844 from a bog deposit in Hackettstown, New Jersey. It cost a considerable sum at the time, $3,000, but Webster figured he could raise the money from private and college donations. Webster, a charming and brilliant professor of chemistry and mineralogy, was never one to let money—or lack of it—stand in his way.

When the mastodon arrived at Harvard in 1846, Webster faced a shortfall. Some men who pledged contributions had failed to pay up. Webster enlisted help from fellow Harvard graduate George Parkman, a wealthy physician, philanthropist, and landlord. Parkman kindly sent around a debt collector on Webster's behalf. Parkman would live to regret his assistance—but not for too much longer.

Even before the mastodon matter, Webster was indebted to Parkman, who, in 1842, had loaned him $400. By 1847, Webster's spendthrift habits had led to deeper trouble. He borrowed an additional $2,000 from Parkman and others, putting up as collateral all his personal property, including his household furniture and his prized collection of minerals. The following year, Parkman learned that Webster had tried to sell that same mineral collection to Parkman's own brother-in-law for $1,200. Outraged at this petty fraud, Parkman began hounding Webster for his money.

On Friday, November 23, 1849, Parkman paid a visit to Webster at his Harvard Medical School office. He was never seen again. A suspicious janitor led investigators to the discovery of Parkman's body, in pieces. They found false teeth and arm and head bones in Webster's furnace. They discovered pelvis and leg bones in a chamber below Webster's laboratory. In Webster's tea chest, they found the thorax and left thigh—as well as the mineral specimen of idocrase pictured here.

Following a sensational trial, Webster was convicted and hanged. As for the mastodon, it remains on display in a glass case, with the original plaque listing the "gentlemen" who contributed to its purchase. On the list was printed—before murder was to link them forever—the names of both Parkman and Webster.

Walcott's Trilobite

Isotelus gigas (eye-soh-TEE-lus), collected between 1871 and 1873.[7]

This trilobite was collected by a remarkable man named Charles Doolittle Walcott. Although he never finished high school, he became a world-class paleontologist and secretary of the Smithsonian Institution. Walcott also discovered the exceptionally important fossils of the Burgess Shale, located in the Canadian Rocky Mountains. (Stephen Jay Gould wrote about these fossils, and Walcott's attempts to classify them, in his best-selling book *Wonderful Life.*)

At age twenty, Walcott gave up his job clerking at a hardware store in hopes of a career in natural history. He moved to Trenton Falls, New York, to work on William Rust's farm and learn from him about the local fossils. Within three years, Walcott had built an impressive fossil collection, digging and preparing specimens in his spare time. In 1873, he sold the entire collection to Louis Agassiz at Harvard's Museum of Comparative Zoology for $3,500, a sizable sum in those days. The collection contained 325 whole trilobites—including the beautifully preserved one shown here—190 crinoids, 6 starfish, and many other specimens, some new to science.

In 1876, Walcott landed his first professional job, becoming assistant to James Hall, state geologist of New York. Hall had a steam-powered saw specially designed for cutting fossils into sections. When Walcott examined thin slices of trilobites, he found evidence that the ancient animals had legs. Until his discovery, no one had had clear proof of how trilobites moved. Walcott's four-page note on trilobite appendages, published in 1876, was a landmark in paleontology.

Trilobites, beloved of scientists and fossil collectors alike, first appeared in ancient seas some 520 million years ago. The entire Paleozoic era is sometimes called the Age of Trilobites, for they were a prevalent life-form for millions of years. Among the first complex creatures on earth, most trilobites had well-developed eyes and a hard exoskeleton. The many thousands of species ranged in length from a few millimeters to more than two feet. Trilobites died out some 248 million years ago, during the mass extinction that marked the end of the Permian period.

The large species pictured here, *Isotelus gigas*, was one of the first trilobites to be described in North America, in 1824. Scientists believe it did not swim well, but rather lay under a thin layer of sediment with only its eyes exposed. If threatened, it could shield its soft underside by rolling into a ball.

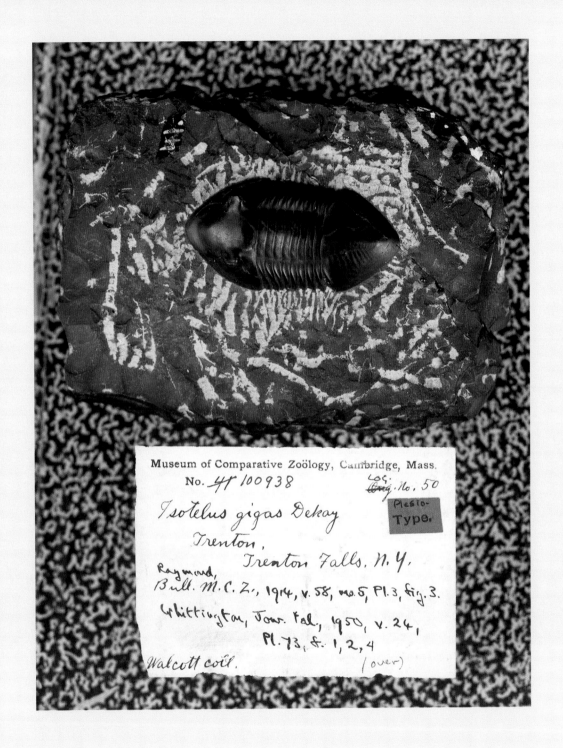

Museum of Comparative Zoölogy, Cambridge, Mass.

No. ~~41~~ 100938   ~~Cat.~~ ~~Cerj.~~ No. 50

Plesio-Type.

Isotelus gigas Dekay

Trenton,

Trenton Falls, N.Y.

Raymond,
Bull. M.C.Z., 1914, v.58, no.5, Pl.3, fig.3.

Whittington, Jour. Pal., 1950, v.24,
Pl.73, f. 1, 2, 4

Walcott coll.   (over)

Frail Child of the Air

Prodryas persephone (proh-DRY-as), collected about 1878.[8]

It seems impossible: a fossil butterfly. Given that so few creatures with bones and teeth survive in the fossil record, how was such a fragile insect preserved? This rarity comes from Florissant, Colorado, where fine shale formed from volcanic ash captured insects in exceptional detail. It was the first fossil butterfly ever found in America.

A woman named Charlotte Hill collected the butterfly, likely at a local ranch. She sent it to Samuel H. Scudder, a Harvard professor who had studied under Louis Agassiz and gone on to become a leading expert on fossil insects. Scudder described the butterfly in 1878, finding it to be "in a wonderful state of preservation, the wings expanded as if in readiness for the [display] cabinet and absolutely perfect, with the exception of the tail of the right hind wing."[9] Beneath the stone, he believed even its tongue was intact.

This butterfly, *Prodryas persephone*, last flew some 35 million years ago on the shores of the ancient Lake Florissant. There, huge sequoias grew, and the climate was likely tropical. The butterfly sipped nectar in peace—until the nearby Thirty-nine Mile volcanoes erupted, spewing out huge clouds of ash and dust. Amidst this upheaval, the butterfly died. As the ash settled, its body sank to the bottom of the lake. Over time, the lakebed ash turned to shale, preserving its body in extraordinary detail, down to individual wing scales.

All told, twelve fossil butterfly species have been found at Florissant. Scudder named eight of them. *Prodryas* bears little resemblance to the modern butterflies of Florissant, where the habitat, high up in the Rockies, is now pine forest and mountain meadow. Instead, the fossil appears more like the tropical *Hypanartia* butterflies of Central and South America.

Scudder made his mark as the author of a three-volume treatise on the butterflies of North America. But he also wrote a charming butterfly guidebook for young people, published in 1895, called *Frail Children of the Air*. At the time, he noted, fewer than twenty fossil butterflies had been found anywhere in the world.

Snapshots from the Jurassic

These ancient sea creatures have been fossilized with near-photographic clarity. They come from the world-renowned Solnhofen limestone formation in Bavaria, Germany. The rocks there reveal an extraordinary panorama of life as it existed some 155 million years ago. It was at Solnhofen that, in 1861, the first skeleton of *Archaeopteryx* was discovered, an extraordinary fossil that suggested an evolutionary link between dinosaurs and birds. And from Solnhofen come many examples of life-forms that rarely survive in the fossil record. These fragile, soft-bodied animals include ancient squids, sea cucumbers, sponges, and jellyfish.

It required unusual conditions for such preservation to occur. Toward the end of the Jurassic period, Solnhofen's shallow lagoons had thick mud and extremely salty water. Scientists believe that during floods or storms dead land animals were swept into the lagoons. With them came sea animals—most still alive—swept in by waves. Eventually, all the bodies settled to the lagoon bottoms, where they were buried in soft mud. So salty was the water that scavengers could not survive in it. Even microbes were scarce. As a result, the animals decomposed slowly, allowing time for minerals to replace soft body tissue.

104220

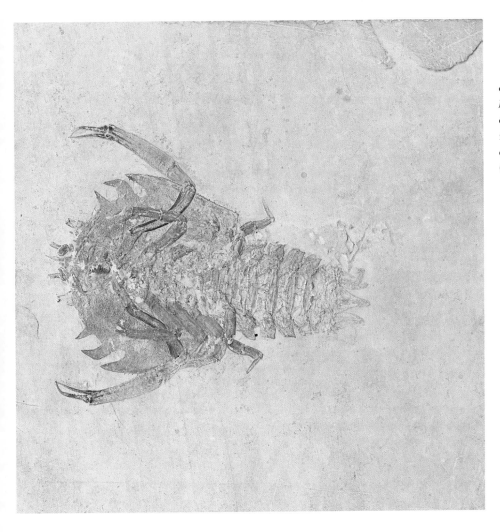

OPPOSITE: Ancient squid *Plesiotheuthis prisca* (plee-see-oh-TOO-this), collected before 1890, and LEFT: Ancient crustacean *Cycleryon propinquus* (sy-CLER-ee-on).[10]

The two fossils pictured here come from Harvard's collection of more than a thousand Solnhofen specimens. The specimen on the right, *Cycleryon propinquus*, is an extinct crustacean, related to modern crabs and lobsters. The specimen on the left is an extinct squid, *Plesiotheuthis prisca*. It is so beautifully preserved that you can see its "pen," or internal shell, and even the hole where its ink sac used to be.

Fine-grained Solnhofen limestone is valued for more than its extraordinary fossils. Since Roman times, the stone has been used for roof and floor tiles. Modern large-scale quarrying began in the early 1800s, after Alois Senefelder invented lithography. From Solnhofen came the lithographic stones used by such artists as Francisco de Goya, Henri de Toulouse-Lautrec, and Marc Chagall.

Blood, Sweat, and Bones

Kronosaurus queenslandicus (kroh-na-SAUR-us), collected in 1932.[11]

In the 1930s, Harvard sent an expedition to Australia, in hopes of collecting specimens of kangaroo, wombat, and Tasmanian devil. The scientists did indeed find marsupials—but they also found something larger.

This skeleton of a gigantic marine reptile was discovered by William Schevill, a graduate student on the expedition who spent months tooling around North Queensland in a dusty Ford pickup, looking for fossils. During his scouting, a sheep rancher named R. W. H. Thomas mentioned that there was something odd protruding from rocks in his paddock. Schevill hurried to the site. Sticking out of the rocks were bones from this enormous pliosaur, *Kronosaurus queenslandicus*.

Because the bones were embedded in solid limestone, Schevill enlisted help from a British migrant trained in the use of explosives. The fellow—nicknamed The Maniac, due to rumors that he had killed a man—dynamited out huge blocks of limestone encasing the fossils. The blocks were then shipped to Harvard, each one weighing some six tons.

Freeing the bones from the limestone presented another time-consuming and expensive task. It took several years to prepare the nine-foot-long skull for public display. The remaining blocks sat in the museum for more than a decade, until Godfrey Lowell Cabot, the Boston carbon-

black magnate, donated $10,000 for the work. (He had a long-standing fascination with sea serpents.)

In the end, the job cost even more. Museum preparators Arnold Lewis and James A. Jenson spent two years extracting the bones, using chisels and acid. The skeleton—about 60 percent complete—was then reconstructed under the guidance of distinguished Harvard paleontologist Alfred S. Romer.

Romer's reconstruction is controversial among today's paleontologists. This is not surprising, given that new generations often reinterpret fossil finds. Unfortunately, Romer made it difficult for scientists to get at the original fossil material, for he encased the real bones in plaster and added fake plaster "bones" where he believed necessary.

Colin McHenry, a pliosaur expert in Queensland, contends that Romer made the *Kronosaurus* too long. He believes that Romer added about eight plaster vertebrae too many, and that the true length should be about thirty feet, not the current forty-two feet. McHenry bases his views on comparisons with related specimens around the world.

Even downsized, *Kronosaurus* would have been a fearsome sight as it patrolled the oceans some 135 million years ago. With teeth the size of bananas, it may well have killed its prey simply by snapping off their heads.

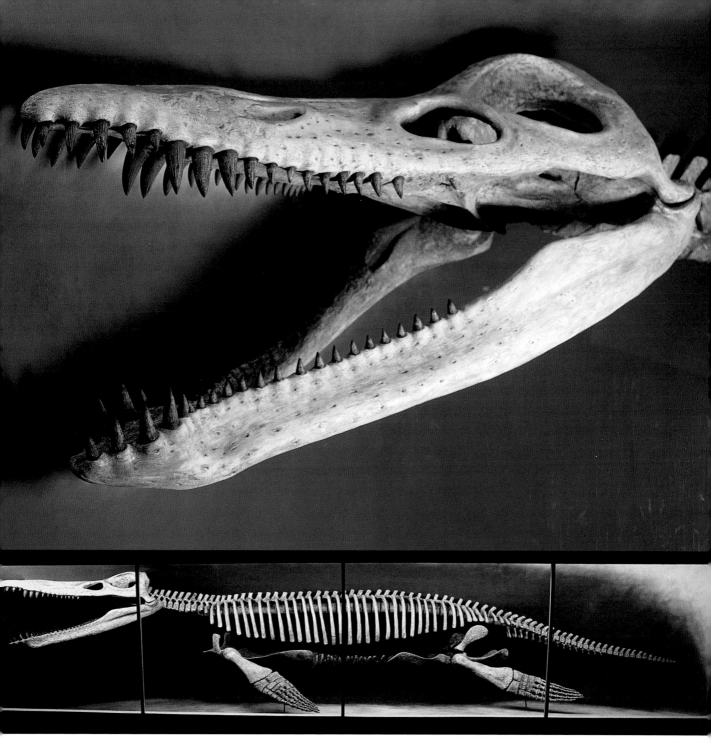

(BOTTOM: Courtesy of the Harvard Museum of Natural History.)

This Is Not a Dinosaur

Dimetrodon milleri (dy-MEE-troh-don), collected in the 1930s.[12]

Alfred Sherwood Romer discovered this *Dimetrodon* skeleton in the red beds of north-central Texas, perhaps his favorite place for bone hunting. In that place, he wrote, "The summer temperatures approximate those of Hell. Almost all the animals and plants bite or sting. Water is scarce and usually unpleasant either before or after taking, or both. But the people of the region are among the finest, and the fossil reptiles to be found there are the world's best."[13]

Romer, a leading twentieth-century paleontologist, had great success on his red-beds collecting trips. From his finds, he named many new species, including this *Dimetrodon milleri*. In 1937, he described it as "an almost perfect skeleton, complete to tip of tail and toes."

Dimetrodon is not a dinosaur. It is a pelycosaur (PEL-i-ca-saur), an early mammal-like reptile. *Dimetrodon* roamed the swamps during the Lower Permian period, about 290 to 256 million years ago, before the age of dinosaurs. During its existence, *Dimetrodon* was likely a top meat-eating predator.

Paleontologists continue to debate the function of the long spines on *Dimetrodon*'s back. Romer believed they supported a sail-like membrane that could catch the morning sun, allowing the reptile to warm up quickly. Others believe the "sail" was used like a peacock's tail, to attract members of the opposite sex.

In 1934, Romer accepted a professorship at Harvard, becoming that rare scholar who excels in research, writing, and institutional leadership. He was the author of many books, including *Vertebrate Paleontology*, the standard textbook on the subject for nearly half a century. He served as director of the Museum of Comparative Zoology from 1946 to 1961 and then, upon retirement, resumed his paleontology research. His wife, Ruth Hibbard Romer, often accompanied him on his fossil-collecting trips; she could, he said, bandage a skull in plaster with the best of them.

When writing for a popular audience, Romer had a light touch. He wrote a delightful account of an expedition to Argentina that contained "good old-fashioned melodrama," including an illegal seizure by the Argentinean police of nearly all the fossils collected. (They were eventually returned to Harvard.) And when, in *The Scientific Monthly*, he wrote of discovering a new fossil amphibian called an *Edops*, he called it simply Grandpa Bumps.

The Very Hungry Dragonfly

Meganeuropsis americana (me-ga-ner-OP-sis), collected in 1940.[14]

Behold the largest complete insect wing ever found. *Meganeuropsis americana* had a wingspan of nearly 2.5 feet, that of a medium-sized hawk. When this fossil was described in 1947, it provided some of the first scientific evidence that insects had once reached gigantic proportions.

Meganeuropsis must have been a formidable predator. The huge dragonfly-like creature would have swooped over steamy coal swamps, snagging smaller insects in its powerful mouth. It had no competition from birds. Not even dinosaurs had yet appeared on earth. *Meganeuropsis* lived during the Permian period, from 290 to 248 million years ago, when the dominant life-form was amphibians.

This spectacular specimen was collected in Noble County, Oklahoma, by Harvard's Frank Morton Carpenter. In 1939, Carpenter had received a batch of Noble County insects from fellow paleontologist Gilbert Raasch. Impressed, Carpenter decided to join Raasch in Oklahoma the following summer. There, in just ten weeks, they collected more than five thousand well-preserved fossil insects, including the one pictured here. Due to World War II, Carpenter had to defer publishing his finds until 1947.

Fossil insects had first captured Carpenter's imagination when he was a teenager, and he never lost interest in them during his long life. As curator of fossil insects at Harvard, Carpenter organized an already vast collection and added greatly to it. In 1992, at the age of ninety, Carpenter made perhaps his greatest contribution to paleoentomology (the study of fossil insects). Drawing on a lifetime of learning, he published an exhaustive two-volume compendium of fossil insects worldwide. The work, part of the series *Treatise on Invertebrate Paleontology*, drew much praise. Carpenter died two years later, soon after receiving a lifetime achievement award from the Entomological Society of America.

As for *Meganeuropsis*, it did not survive the era. The Permian ended in the greatest mass extinction ever known, with about 70 percent of all land creatures dying out. Some scientists contend that the cause was a comet or asteroid crashing into Earth. About 183 million years later, a similar disaster likely led to the extinction of the dinosaurs.

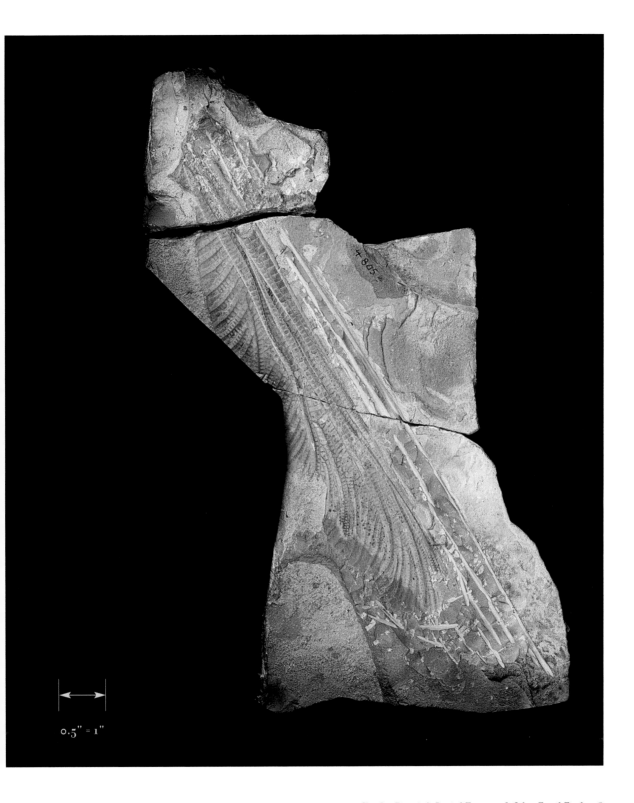

0.5" = 1"

The Astonishing Turtle

Stupendemys geographicus (stoo-pen-'DE-meez), collected in 1972.[15]

This shell is one for the record books. It belongs to the largest species of turtle ever discovered, *Stupendemys geographicus*. You can even find its picture in a Guinness book of world records, *The Guinness Book of Animal Facts & Feats*.

The shell—seven feet two inches long—was one of two found in the Venezuelan desert in 1972, during a joint expedition conducted by Harvard and the Universidad Central de Venezuela. A paleontology student spotted part of the shell poking out of a hillside in Urumaco, a place of cactus and thorn scrub.

With picks and shovels, expedition members dug out the shell and broke it into about thirty pieces for shipment. At Harvard's Museum of Comparative Zoology, chief preparator Arnold Lewis assembled the pieces like a jigsaw puzzle, using plaster and fiberglass to fill in the gaps. The expedition's turtle expert, Roger C. Wood of Stockton State College, gave it the name *Stupendemys*, meaning "astonishing turtle."

The second *Stupendemys* shell from 1972, although less complete, turned out to be even longer, at seven feet seven inches. As the type specimen, it became the property of the Museo de Ciencias in Caracas, Venezuela. By prior agreement, that museum was to receive all the type specimens—those chosen to represent new species—as well as half of all other specimens.

Stupendemys lived about 6 million years ago, probably in a system of freshwater rivers and lakes belonging to the early Amazon Basin. With it swam enormous crocodiles and other species of large turtles. (The Urumaco site has also yielded up fossil remains of the largest rodent that ever lived, *Phoberomys pattersoni*, which weighed about 1,500 pounds.)

Interestingly, *Stupendemys* was a side-necked turtle, one that pulled its head sideways under its shell. Today, most turtles belong to a second group, the hidden-necked turtles, which retract their heads straight back into the shell. In South America, however, at least twenty-five species of side-necked turtles remain in existence.

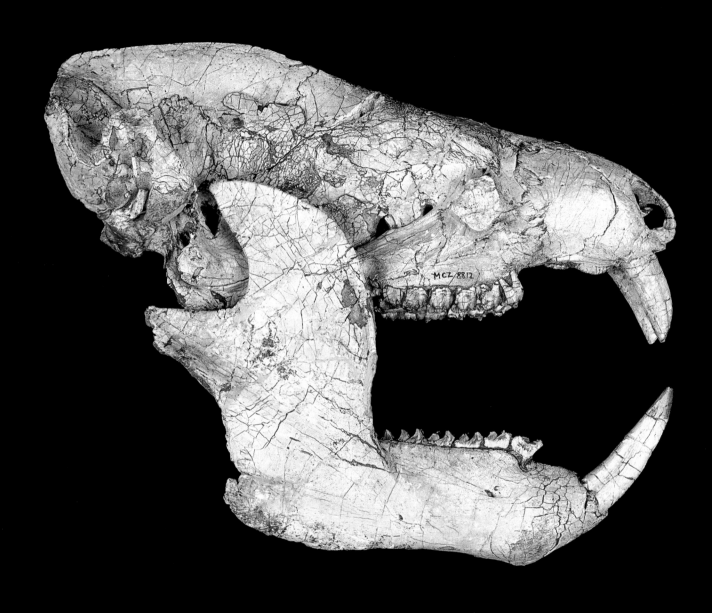

Perfect after 180 Million Years

Skull of *Kayentatherium wellesi* (kay-ent-a-THEER-ee-um), collected in 1981.[16]

This skull belongs to the world's finest specimen of *Kayentatherium wellesi,* an early mammal-like reptile that lived during the Lower Jurassic, about 180 million years ago.

Kayentatherium belongs to a fascinating group of extinct animals called cynodonts (CY-noh-donts, "dog teeth"). Scientists believe that certain cynodonts represent the evolutionary link between reptiles and mammals. Although this particular species was not a direct mammal ancestor, its teeth and jaws reveal certain mammal-like characteristics.

This beautifully preserved specimen was discovered by Charles Schaff in the summer of 1981, on Navajo Nation land east of Flagstaff, amidst the stunning red rock cliffs of the Painted Desert. Most of the year, Schaff sorts and prepares fossil specimens at the Museum of Comparative Zoology. But come summertime, he packs up his tool kit and goes off to hunt for bones, working with distinguished Harvard paleontologist Farish Jenkins Jr.

Trudging through the Arizona desert in 1981, Schaff scouted the flat areas first, trying to avoid any unnecessary climbing in the 110-degree heat. He was not exactly traveling light, carrying such basic equipment as chisels, hammers, brushes, plastic bags, canteens, lunch, a hand lens—and the all-important toilet paper, used for wrapping fossils.

Initially, Schaff spotted fossil bone fragments that had washed down from the cliffs. He traced the fragments up to a hillock, where he found more bone. As he began removing rock, he knew he had found something extraordinary. All the animal's ribs were articulated—still joined—after 180 million years. The skull was complete, as were the vertebrae. Schaff found the entire animal intact, except for the back legs and the pelvis. Clearly, the animal had not traveled far from where it died, as, he said, he found it "all laid out in a nice relaxed pose."[17]

The teeth were perfectly preserved. The *Kayentatherium*'s large fanglike incisors may have been used for digging or defense. The highly specialized back teeth suggest that the ancient animal was an herbivore and may have had a diet similar to that of modern rodents.

3 Seven-Colored Tanagers
& Other Emblems of Biodiversity

Vampire bat, *Vespertilio vampyrus*. At night sucks in the blood of the sleeping, the combs of cocks, and the juice of palm trees.

—Carl Linnaeus, *Systema Naturae*[1]

Linnaeus described an astounding number of species during his lifetime, more than 9,000 plants, 2,100 insects, and 470 fish—as well as vampire bats and many other mammals. Despite this Herculean effort, by his death in 1778 Linnaeus had cataloged only a tiny fraction of the earth's biological diversity.

A century later, scientists grew more ambitious in their quest to survey the world's species. Biologists described thousands upon thousands of life-forms, as travelers explored ever more remote regions of the planet. Although efforts slowed somewhat in the twentieth century, by the year 2000 biologists had assigned names to more than 1.5 million species of animals and plants. No precise count is possible because no central inventory of known species exists. Yet scientists have completed databases for several groups, including birds and mammals, and their surveys grow more sophisticated with every passing year.

The specimens selected for this chapter reflect just a small sampling of the biodiversity represented in Harvard's natural history collections. Some specimens were chosen because they are beautiful—such as the tanagers and the regal lily. Others were picked because they have interesting life histories, such as the tuatara. The shipworms had an intriguing link to the *Titanic*. As for the "fanny shell," it was selected simply because of its scientific name.

Also, this chapter, more than the others, allowed photographer Mark Sloan to select specimens based on their visual appeal. Armadillos are a particular favorite of his, as are the weevils, in their fantastic variety of color and form.

Now, more than two hundred years after Linnaeus, biologists have once again launched a great effort to catalog all the earth's species, even as many plants and animals are disappearing. The vast majority of undiscovered life-forms are inconspicuous—insects and other arthropods, fungi, microorganisms. There are likely more than 4 million species of insects in the world, compared to roughly 4,600 species of mammals. Yet new birds and fish are added to the ranks

Ermines, showing seasonal color changes. In winter, their coats are pure white, except for the tips of their tails.

nearly every year, and even a new mammal is occasionally found. Although scientists disagree on the total number of species on earth, with estimates ranging widely, there is some tentative convergence on the number 14 million.

As the world's catalog of species grows, so, too, will the collections of the world's great natural history museums. They may never contain the totality of the earth's biodiversity, but then even Linnaeus had to recognize the limitations of his endeavors.

The Regal Lily

Lilium regale (LIL-ee-um), collected in 1908.[2]

The great plant hunter Ernest Henry Wilson introduced many Chinese plant species to the West, including the dove tree and the kiwi. But he is perhaps best remembered for this, the regal lily. With its huge white trumpet flowers and heady fragrance, the regal lily has long been a garden favorite. Yet until Wilson's 1910 expedition—which nearly cost him his life—the plant grew only in the rugged mountains of western China.

The English-born Wilson was a veteran of three successful trips to China to collect plants. In 1909, Charles Sprague Sargent, director of Harvard's Arnold Arboretum, convinced Wilson to undertake yet another expedition. Leaving Boston in March of 1910, Wilson crossed to Europe and rode the Trans-Siberian Railway to Beijing. From there, he made his way to Songpan, in Sichuan Province, arriving toward the end of August. Although Sargent had asked Wilson to collect conifer seeds, Wilson had a different goal in mind. His primary quest was the regal lily, which he had encountered a few years before. "I was determined that it should grace the gardens of the Western world," he wrote. "That such a rare jewel should have its home in so remote and arid a region of the world seemed like a joke on nature's part."[3]

To reach the lilies, Wilson had to descend the Min River gorge, following a treacherous road trafficked by mule trains transporting tea and cotton cloth. On the eighth day of his descent, fighting fierce winds, he made camp. In this unlikely place, the *Lilium regale* grew, "here and there in abundance on the well-nigh stark slate and mudstone cliffs." He spent several days arranging for local workers to dig about six thousand bulbs and transport them for shipment to Boston. With plans completed, Wilson and his entourage headed for the comforts of Chengdu, the Sichuan capital.

They made good progress until startled by a sudden rockslide. A boulder hit Wilson's right leg, breaking the bones in two places. Using Wilson's camera tripod as a splint, his companions carried Wilson, in agony, to Chengdu, rushing to arrive in just three days. There, a missionary doctor treated the break, but infection had set in, and it appeared the leg would have to be amputated. Finally, a French army surgeon cured the infection, and Wilson's leg began to knit. Ever afterward, his leg was sound, but crooked and almost an inch short, leaving him with a permanent "lily limp."

The precious regal lily bulbs, encased in clay and packed in charcoal, arrived safely in Boston a few days after Wilson, in the spring of 1911. The first stock was established in Farquhar nursery in Roslindale, Massachusetts, where the lily thrived. "Each year, it adds to the pleasure of millions of folk," Wilson wrote. "The price I paid has been stated. The regal lily is worth it and more."

No. 1446. ARNOLD ARBORETUM.
EXPEDITION TO CHINA, 1907-09.
Western Szechuan.

TYPE *Lilium regale* Wilson

2-4 ft. Fl purple median-golden inside. Min Valley
Coll. E. H. Wilson. Mao-chou. Alt. 2,500-4000ft.
6/08

Type

GRAY HERBARIUM
HARVARD
UNIVERSITY

THE HARVARD UNIVERSITY HERBARIA
00029973

They Dined on the *Titanic*

Xylophaga muraokai (zy-loh-FAY-ga), collected in 1965; wood sample with bore holes; toy model of the Titanic.[4]

When the wreck of the *Titanic* was discovered in 1985, scientists wondered why so little of its woodwork remained. Ruth Dixon Turner solved the mystery. Turner, a world authority on shipworms, had proven that certain species of shipworm exist at great depths. Some 2.5 miles below the surface, these "termites of the sea" had devoured the *Titanic*'s elaborate woodwork, leaving spongy masses filled with bore holes.[5] Not even the hard teak decks were spared.

Turner was a pioneer in many endeavors. In 1971, she became the first woman scientist to dive in *Alvin*, the deep-sea submersible owned by the Woods Hole Oceanographic Institution on Cape Cod. She went on to do more than fifty *Alvin* dives, conducting research on deep-sea shipworms. In 1976, she became one of the first women to receive tenure in the sciences at Harvard.

Shipworms, her specialty, get their name from their long, wormlike bodies. But in fact they are mollusks, not worms. Their small, clamlike shells sit on their heads like helmets. The edges of these modified shells are razor-sharp, perfectly adapted for boring into wood.

Shipworms invade new wood as tiny larvae, leaving entrance holes that may be nearly invisible. Inside the wood, the larvae grow into adults up to several feet long. They tunnel through the wood, eating as they go. Often, shipworm damage goes unnoticed until the wood becomes so laced with holes that it disintegrates. Shipworms have destroyed countless wooden boats, piers, and docks. In 1502, during his fourth voyage to the Caribbean, Christopher Columbus had to abandon his ships due to shipworm damage.

To learn more about shipworms, Turner lowered wood samples into the sea and suspended them at various depths. Then, months or years later, she would retrieve the samples and examine the shipworms that had bored into them.

The specimens pictured here, in the glass vial, were named and described by Turner. These *Xylophaga muraokai* came from a wood sample that Turner left at a depth of fifty-six hundred feet for two years, off San Miguel Island in California. Quite likely, the shipworms that ate the wood of the *Titanic* belonged to the same genus.

The Fanny Shell

Hyridella fannyae (hy-ri-DEL-a), collected in 1957.[6]

Zoologists usually give sober names to newly discovered animals. Typically, they choose names—based on Latin or Greek—that refer to the animal's color, shape, or place of discovery. Some taxonomists do, however, have a sense of humor about their work. Among them is Richard I. Johnson, a gentleman scholar affiliated with Harvard's mollusk department for more than sixty years.

The shapely mussel pictured here, native to New Guinea, bears the name *Hyridella fannyae*. Johnson described the species in 1948, noting that it was "appropriately named" for his sweetheart at the time, Fanny Day Farwell. The mussel, he wrote, "may be distinguished from the other members of the genus by the remarkable development of the posterior ridge and the consequent swelling of the post-basal region."[7]

In 1965, Johnson named an American freshwater mussel after his wife, Marjory (Peggy) Weld Austin, calling it *Anodonta peggyae*. Following their divorce, in 1983 he named a new species of pearl-shell mussel *Margaritifera marrianae* after his second wife, Marrian Geer Gleason.

Johnson is not alone in being a taxonomist with a wink. On April Fools' Day, 1993, Arnold S. Menke of the U.S. National Museum published a long list of "funny or curious zoological names," all of them real. In the beetle genus *Agra*, one is named *Agra vation*, and a braconid wasp is named *Heerz lukenatcha*. There are trilobites named for Rolling Stones members, *Aegrotocatellus jaggeri* and *Perirehaedulus richardsi*. And a species of owl louse, *Strigiphilus garylarsoni*, honors the cartoonist of *The Far Side*.

Additionally, there is a fossil snake of the genus *Montypythonoides*, and a wasp named *Polemistus chewbacca*, after the *Star Wars* character. The fly *Dicrotendipes thanatogratus* honors the band Grateful Dead, and rock star Sting has a tree frog named after him, *Hyla stingi*. Linnaeus himself was not above naming a lowly weed *Siegesbeckia*, to insult botanist Johann Siegesbeck, who had attacked Linnaeus's ideas on plant classification.

True Attachment

Linophryne bicornis (lye-noh-FRY-nee), collected in 1995.[8]

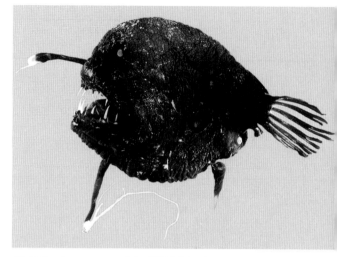

(C. P. Kenaley, courtesy of the MCZ Ichthyology Department.)

This is an anglerfish, one of the oddest fish in the sea. Actually, two anglerfish appear in the photograph. The obvious one is the female. The small, thumblike appendage dangling from her abdomen is the male. The male is not just visiting, but rather has become permanently attached to the female, joined by outgrowths from its snout and lower jaw.[9]

Anglerfish inhabit the deep sea, where a "parasitic male" makes good evolutionary sense. In the dark and barren depths, it is hard to find a match! Male anglerfish generally cannot feed themselves, but they are good swimmers with highly developed smell receptors. When a female drifts by, the male catches her scent and hooks onto her skin.

In his book *Promiscuity*, Tim Birkhead writes that anglerfish are something of an exception among fish, for they are mainly monogamous: "Male anglerfish permanently attach themselves to the underside of a female from an early stage in life. Little more than bags of sperm, males extract nutrients by fusing their tissues directly with those of the female and tapping directly into her blood supply."[10] The male thus benefits by receiving "free lunches for life." Also, given that most females have just one parasitic male attached, the male stands a good chance of being the father of her offspring. As for the female, she benefits by receiving a reliable source of sperm for her eggs.

Anglerfish are remarkably diverse, comprising some 150 species. This pair belongs to the species *Linophryne bicornis*, first described in 1927. The two were collected off the southern New England coast in 1995 by a Harvard researcher, at a depth of some three thousand feet below sea level. The female measures about four inches, while the male is just three-quarters of an inch long.

Another interesting feature of anglerfish is that the females have bioluminescent lures. In the photograph, two lures can be seen, one projecting off the head and the other off the chin, both with white "bulbs" at the tips. The bulbs contain light-producing bacteria that attract prey to the sedentary female, who simply waits for the food to come to her.

Of Tanagers and Blue Skies

Tanagers are breathtaking. Nearly all live in the neotropics—southern Mexico, Central and South America, and the West Indies—where they frequent forests and gardens, plucking berries or hunting for insects.

Five of the birds pictured here belong to the genus *Tangara*. This is the largest tanager genus, with about fifty species. Sometimes called callistes—from the Greek *kalli*, meaning "beautiful"—they travel in small flocks through the forests.

Even among the callistes, one species stands out for its extraordinary plumage. This is the seven-colored tanager, an endangered species. Two appear on p.86: one in the top row, second from right, with the yellow patch; the other in the bottom row, on the left, with the blue and purple belly. Look closely, and you can detect all seven colors.

Blue feathers are particularly fascinating, in terms of their physics. Blue feathers do not contain a blue pigment. Rather, they appear blue because of the way that light bounces off them. For many years, blue bird feathers were thought to look blue for the same reason that the sky looks blue. This phenomenon is called the Tyndall effect, after John Tyndall, a nineteenth-century English physicist. He discovered that when a beam of light scatters off tiny particles, the color blue appears. This is because blue and violet light are the most prone to scattering, as they have the shortest wavelengths of all colors. The sky appears blue, Tyndall determined, because blue light bounces off the tiny impurities in the atmosphere.

More recently, however, an expert in bird coloration has argued that this explanation does not apply to bird feathers. Richard O. Prum and his colleagues at the University of Kansas believe their data shows that blue feathers look blue for the same reason that oil slicks appear colorful. This effect is due to interference, not to Tyndall scattering. In an oil slick, iridescent colors appear because of differences in the distance that light waves travel when reflected off its surface. Within feathers, tiny air bubbles may be aligned so that one particular color—such as blue—is brilliantly reinforced.

Prum first found interference at work in bird feathers, but he has since identified it in the blue of bird skin, mammal skin, frog skin, and dragonflies. Indeed, he believes he has overturned the Tyndall effect nearly everywhere it was thought to occur in biology, except in the iris of the human blue eye.

OVERLEAF: Top row, left to right: *Tangara nigroviridis, Chlorophanes spiza, Poecilothraupis ignicrissa, Tangara fastuosa, Tangara mexicana.* Bottom row, left to right: *Tangara fastuosa, Euphonia musica, Tangara xanthogastra, Ramphocelus sanguinolentus.* Various dates.[11]

Catalogue Number.	Original Number and Set Marks.	Name.	Taken at or near	When Taken.	
6126	676 ⅖	*...s motacilla*	Middles...	...nnecticut	May 8, 1896	Claren...
127	676 ⅙	"		"	" "	
128	676 ⅐	"		"	" 9, "	
129	676 ⅗	"		"	" 10, "	
30	676 ⁴⁄	"		"	" 10, "	
31	676 ⅗	"nest discarded"		"	" " "	
32	676 ⅗	"		"	" 13, 1898	
33	676 ⅕	"		"	" 19, "	
34	676 ⅖	"		"	" 24, "	
35	676	"		"	" 25, "	
6	676 ⅘	"		"	"	
7	676 ⁶⁄	"		"	"	
	676 ⅞	"		"	"	
	676 ⅞	*...pis*		"	"	
	681 ⅕			"	"	
	681 ⅕			"	"	
	681 ⅖			"	"	
	683 ⅕	*Icteria vir...*		"	" 1891	
	684 ¼	*Sylvania m...*		"	" 1, " "	
	684 ¼	"		"	" , 1898	
	687 ¼	*Setophaga ...tica...*		"	May 28, 1891	
	624 ¼	*Vireo olivaceus*		"	June , 1888	
	624 ⁴⁵⁄₁₆ [2+4 eggs]	"		"	" 14, 1893	
	628 ¼	" *flavifrons*		"	" 1, 1890	
	631 ¼	" *...noveboracen...*		"	" 1889	

7/90
Tangara nigroviridis berlepschi
Chaupe; D. Cajamarca,
6000'. July 26. Peru Exped. 1933
M. A. Carriker, Jr.
Acad. Nat. Sci. 116172
Philadelphia

No. 19253F.
Tangara ...
Loc. Brazil, ...
— MUSEUM COMPARATIVE ZOOL...

W. Brewster,
Original No.

Identification.	Eggs.	Position of Nest and Remarks.

Sure | Fresh | Of dead t... ...er nest of moss & rootlets placed ... roots of a
 | | fallen
 " | Inc. well begun | Of dece... ...nest of moss & rootlets placed ... roots of a
 " | Inc. slight | Of de... ...k, the nest over the we... ...ner nest of bits of moss ...ed in a ban
 " | Fresh | Of de... ...st contained 2 Cowb... ...otacilla].
 | | near... ...ed with moss & rootlets ... low bank
 ...reets & eggs | " | Of de... ...placed in a bank near t...
 ...known
 ...re | " | Of dece... ...lining of bark, grass & hai... ...ng roots
 | | a fall... ...the water.
 " | | Of decea... ...lined with ... grass a... ...ed amon
 | | roots... ... maple tr... a bro...
 | | Among roo... ...overturned tree ... water, ha... ...leaves, gras
 | | & rootlets ... with fine ... ...ootlets.
 ...ll known & | Inc. slight | Among r... ...fallen tree ov... ...composed ...leaves, lin
 ...d from nest | | with fin... ...and rootlet...
 ...s seen | Fresh | In very... ...tion, on a ledg... ...leaves ... near
 ...g about | | Nest of a... ...ined with f... ...d hors...
 ...nests & eggs | Inc. | Under a... ...old cart... ...s. from...
 ...e known | | leaves, line... ...grasses &...
 ...well known & | Fre... | ...mong root... ...fallen tree ... nest of...
 ...hed from nest | | ...th fine gr... ...d hor...
 Sure | | ...mong roots u... ...bank ... ...t of dead leaves.
 | | ...fine roots... ...the h...
 | | ...tion unusual... ...ge... ...eaves, about 6 ft. up.
 | | ...of leaves a... ...lets.
 Sure | | ...d leaves, strip... ...ned with finer grasse
 | | ...edar bush in...
 " | | ...ves, strips of bark... ...asses, lined with fir
 | | ...s, placed in bus... ...tu...
 | | ...d leaves, weeds and coa... ...ned with fine grasse
 | | ... tree two feet up in pa...
 | | ...ad leaves, coarse grasse... ... lined with fine gra
 | | ...d in low bush in pa...
 | | ...ow bush, nest composed ... ...strips of coarse gr
 | | ...k, lined with fine gra... ...in woods.
 | | ...dead leaves, strips of bark... ...spiders' webs, lined
 Bird seen on nest | | ...he hair like rootlets, pl... ...laurel bush 2 f
 Sure | | ...f strips of plants, barks &... ...fibres, lined with h
 | | ...apple tree about 20 ft. up...
 | | ...f fibres, strips of bark &... ...ned with fine gras
 | | ...in the horizontal fork of...
 | | ...f strips of weeds and gr... ...lined with fine
 | | ...placed in horizontal fork of... ...contained 2 Vireo's & 4 C
 | | ...f strips of bark, weeds & gra... ...ith lichens and wit
 | | ...the tree, nest placed in for... ...y 40 ft. above the gr
 | | ...f strips of weeds, bark s... ...os & decayed wood, w
 | | ...a forked limb of a lo... ...h.

Specimen label: Coll. of E. & O. Bangs, No... SURINAM, PARAMARIBO, AUTUMN

Specimen label: COLLECTION OF W. W. BROWN ... Collected by W. W. Brown

Specimen label: M.C.Z. 26820... Poecila Maregas...

Specimen label: ...PERUVIAN EXPEDITION 1939...

Specimen label: Xanthogaster 256 696

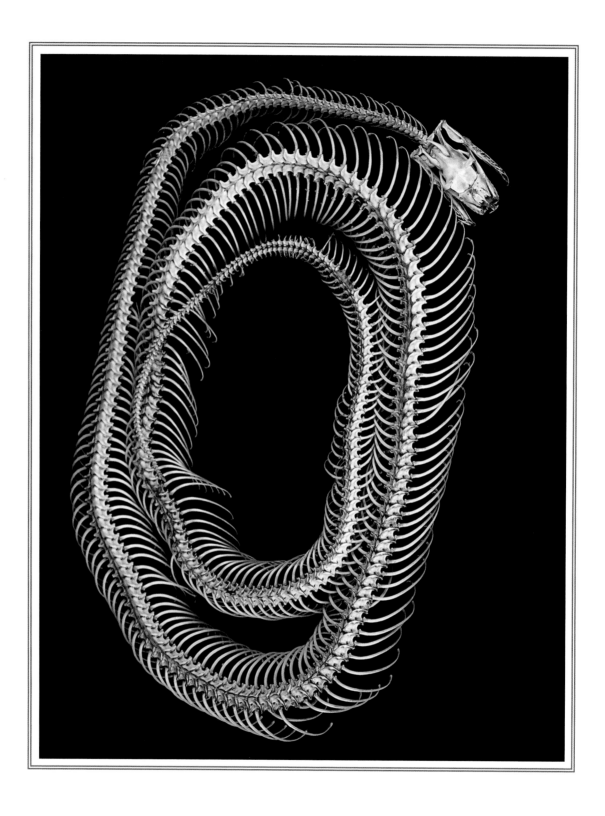

Boa Bones

Skeleton of *Boa constrictor imperator* (BOH-ah), donated in 1992.[12]

Natural history museums receive specimens from many sources. This skeleton belonged to a pet boa constrictor, donated to the Museum of Comparative Zoology in 1992 by Marla Isaac, a Massachusetts reptile enthusiast who specializes in wildlife education.

When the boa died, Isaac put it in her freezer. Then she called José Rosado, longtime collections manager in herpetology, who accepted the specimen for Harvard's museum. The snake's flesh was removed to expose its exquisite skeleton, with some three hundred vertebrae. Although the specimen has limited scientific value—it was likely bred in captivity in the United States—it is nonetheless useful for anatomical study.

Isaac purchased her first boa during college. She went on to breed boas, fascinated by the variations in their diamond and oval skin patterns. Interestingly, boa constrictors give birth to live young, with twenty to fifty typically born at one time. At birth, they measure about eighteen inches long. They continue to grow throughout their lives, reaching an average length of some seven feet and a weight of about sixty pounds.

Boas are carnivores. They grab an animal in their sharp teeth, then coil around it and squeeze until it suffocates. Like all snakes, boas swallow their prey whole. In the wild, they feed on birds, lizards, rodents, and even small pigs.

Pythons and boas belong to the same family, called Boidae. The family includes some of the largest snakes in the world, including the green anaconda, a species of boa that can grow to thirty feet and more than 300 pounds. Some species in the Boidae family have temperature-sensitive scales—called pits—around their mouths. These act like heat detectors, to help the snakes locate warm-blooded prey in dense foliage and darkness. Like all snakes, boas "smell" with their tongues, picking up scent chemicals from the air.

Boas are found mainly in the New World, although species exist on Madagascar and the Solomon Islands. The subspecies pictured here, *Boa constrictor imperator,* is native to Central and South America.

Worldwide Weevils

A theologian once asked the twentieth-century British biochemist and writer J. B. S. Haldane what could be inferred about the nature of God from studying creation. Haldane reportedly replied, "That He has an inordinate fondness for beetles."

Beetles are the champions of biodiversity. Of the earth's known animals, fully one-fourth are beetles; some 350,000 species have been named, and thousands of new species are collected every year. They can be found in virtually every habitat, short of salt water and polar ice. Their small size, tough exterior, and ability to fly have contributed to their success, as has their adaptability in finding food. Some beetles are plant-eaters, while others are predators, parasites, or scavengers.

Among beetles, the largest group is weevils, with some fifty thousand known species. More are being discovered all the time, making weevils the largest family in the animal kingdom. Weevils can easily be distinguished by their long rostrum, or snout. At the tip are tiny chewing mouthparts that can do

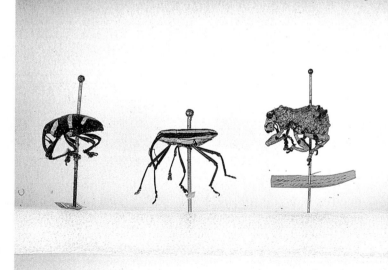

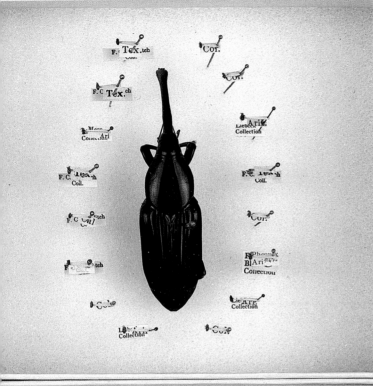

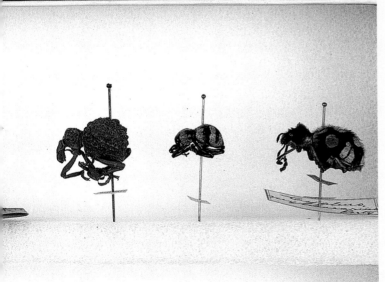

Curculionidae species (cur-coo-lee-OH-ni-dee), various dates.[13]

considerable damage. Some weevils drill holes in nuts and seeds, into which the females deposit their eggs. Boll weevils and many other members of this family are notorious destroyers of crops and grains.

Increasingly, weevils are appearing on the Internet. Harvard entomologists have taken a leading role in putting their collections online, as a tool for researchers. The Museum of Comparative Zoology holds thirty-three thousand insect type specimens, those important individuals chosen to represent a new species. One by one, they are being digitally photographed and entered into a database.

Until recently, weevil researchers had to spend time and money traveling to natural history museums to view type specimens. Alternatively, the specimens were shipped to researchers, risking damage or loss. But online weevils have significant advantages, for they can easily be located, viewed, and compared. Harvard's entomologists are hoping their database will prompt new and important studies and contribute to scientists' understanding of the world's insect biodiversity.

A Reptile Survivor

Mount of *Sphenodon* species (SFEE-noh-don), no date.[14]

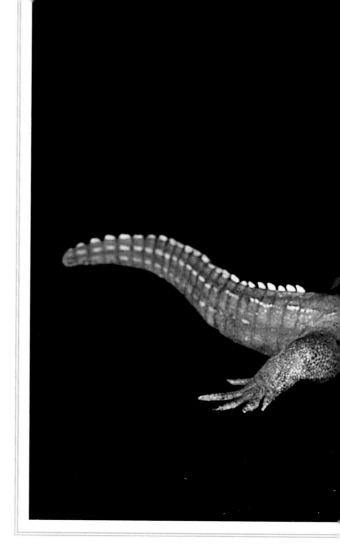

Contrary to appearances, tuataras (too-ah-TAR-ahs) are not lizards. In fact, they are quite distinct from all other living reptiles. Since the time of the dinosaurs, tuataras have survived virtually unchanged, the last remnant of an ancient reptilian order. As for their unusual name, it means "spiny back" in Maori, the language of New Zealand's indigenous people.

Tuataras and their relatives were once quite widespread. They declined with the rise of mammals, after the extinction of the dinosaurs about 65 million years ago. Eventually, tuataras died out everywhere except New Zealand, a place that—until the arrival of humans—had no land mammals. The introduction of the kiore, or Polynesian rat, was devastating for tuataras. They survived on the mainland until the late 1800s, but now remain only on offshore islands, where they are protected.

Scientists once believed that the tuatara was a lizard. But in 1867, Albert Gunther determined otherwise. Gunther, a curator at the British Museum in London, examined a tuatara preserved in alcohol and linked it to the order of reptiles called *Rhynchocephalia* ("beakheads"). The order flourished during the Mesozoic era, which lasted from the time of the earliest dinosaurs, 248 million years ago, until 65 million years ago.

Tuataras reproduce extremely slowly, making it difficult for diminished populations to recover.

Sexual maturity does not occur until they are between fifteen and twenty years old. Egg formation requires up to four years in a female, longer than for any other reptile. And the eggs take some twelve months to hatch.

Additionally, global warming could prove disastrous for tuataras. The sex of tuatara eggs is determined not by sperm, but rather by temperature. Biologists in New Zealand have shown that tuatara eggs incubated at twenty-one degrees centigrade

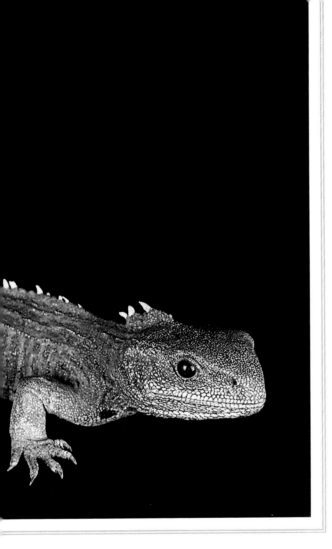

produce hatchlings that are evenly divided between males and females. At one degree warmer, nearly all the hatchlings are male. At one degree cooler, nearly all are female. The research was led by tuatara expert Charles Daugherty at Victoria University of Wellington, New Zealand. He has been at the forefront of efforts to restore tuatara populations to sustainable levels.

OVERLEAF: Left, *Dasypus novemcinctus* (DAS-ee-pus); middle, *Chaetophractus nationi* (ky-to-FRAC-tus); right, *Tolypeutes tricinctus* (tol-ee-POO-teez). Armadillo skin, *Euphractus sexcinctus* (yoo-FRAC-tus). Various dates.[15]

Armadillos in Four Forms

These armadillos reside among the eighty thousand specimens in Harvard's mammal collection. Armadillos are fascinating creatures, with unusual reproductive lives. Through a phenomenon called polyembryony, several species of armadillo always give birth to four identical young, all of the same sex.

On the left is a nine-banded armadillo, the state small mammal of Texas. The armadillo in the middle, the Andean hairy armadillo, is unusual in having hair growing between its scales. On the right is a Brazilian three-banded armadillo, preserved in its closed-up state. Only two species of armadillo can roll themselves in a ball this way, folding head, legs, and tail into the shell.

The flat armadillo skin is that of a six-banded armadillo, a species that digs its burrows in the savannas of South America.

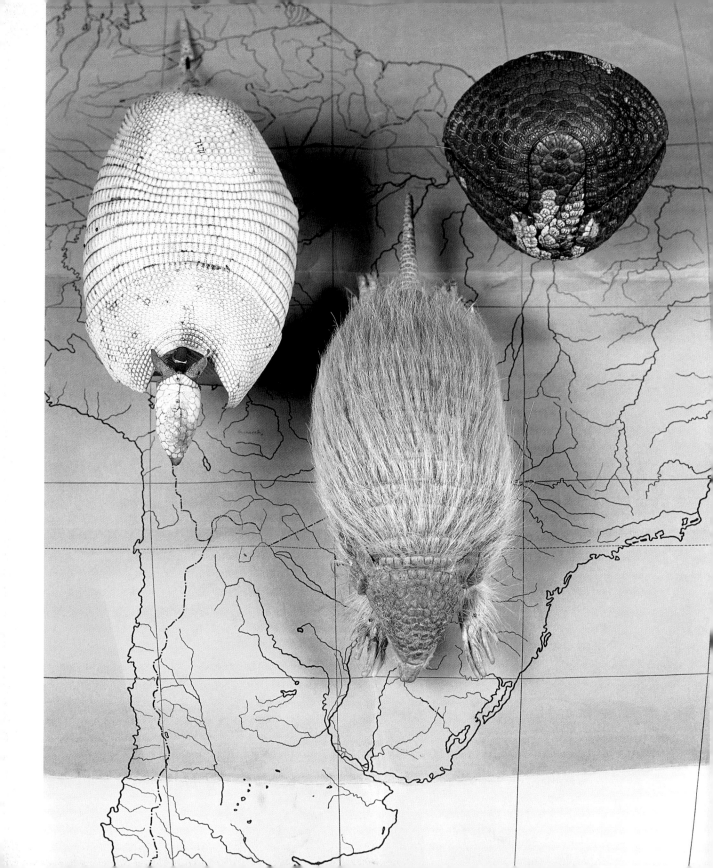

Double Identity

Morpho species (MOR-foh), no date.[16]

This morpho butterfly is male on one side, female on the other. Scientists call such rarities gynandromorphs (ji-NAN-dro-morphs). The male—typically more brilliant than the female—is on the left side, with less black on its wings.

Gynandromorphs result from a genetic aberration, the cause of which is not clearly understood. The word *gynandromorph* comes from three Greek roots: *gyn* (woman), *andro* (man), and *morph* (form). Gynandromorphs are unable to reproduce.

Butterflies are not the only animals in which the phenomenon occurs. Gynandromorphs can also be found among birds, spiders, beetles, lobsters, mosquitoes, and other bilaterally symmetrical animals. Among butterflies, the most dramatic specimens—like this one—display obvious differences between the male and female wings. Such specimens have long been prized by collectors.

In Vladimir Nabokov's autobiography, *Speak, Memory*, he describes the painful loss of a gynandromorph butterfly he had captured himself. His stout Swiss governess accidentally sat upon a tray of choice specimens, leaving them squashed and mangled. He writes of how one butterfly might be mended with glue, but "a precious gynandromorph, left side male, right side female, whose abdomen could not be traced and whose wings had come off, was lost forever: one might reattach the wings but one could not prove that all four belonged to that headless thorax on its bent pin."[17]

The term *gynandromorph* applies only to animals. It is not to be confused with the term *hermaphrodite*. Animal hermaphrodites have—as a natural biological condition—both male and female reproductive organs. Banana slugs are hermaphrodites that alternate sex roles during their lengthy mating ritual. Earthworms are hermaphrodites as well.

Dodos, Real and Fake

Model of dodo by Rowland Ward, made about 1900; skeleton of *Raphus cucullatus* (RAY-fus), acquired in the 1870s.[2]

The dodo, that great symbol of extinction, died out sometime before 1700. Fat and flightless, the species could have evolved only on an island. For, as David Quammen writes in *The Song of the Dodo*, islands are "havens and breeding grounds for the unique and anomalous. They are natural laboratories of extravagant evolutionary experimentation."[3]

The island that harbored the dodo was Mauritius, located in the Indian Ocean not far from Madagascar. Predators were few. That is, until Europeans arrived with pigs and monkeys, which devoured dodo eggs laid in nests on the ground. Dodo populations were also decimated by hungry sailors, who hunted the birds and salted them down for provisions. By the 1640s, dodos were already rare. The final date of disappearance is a matter of debate. There was a sighting in 1681, though some ornithologists believe the birds seen were actually red rails. The last dodo may well have expired in the 1660s.

Sad to note, the dodo's closest relative is also extinct. The Rodrigues solitaire (*Pezophaps solitaria*) was another large and flightless member of the pigeon family. It, too, had inhabited an island, near Mauritius, before dying out about 1765. British scientists recently determined the closeness of the two species by analyzing DNA extracted from museum specimens. The dodo DNA used for the study came from the world's only specimen containing soft tissue. This is the "Alice in Wonderland dodo," held by the Oxford University Museum of Natural History in England, so called because it inspired the Lewis Carroll character. The head and right foot are all that remain.

Harvard's dodo skeleton, though a composite, is a rarity of considerable scientific value. The skeleton contains both real bones and some plaster reconstructions, which appear smoother than the genuine ones.

As for the dodo model, it is a fake and of no scientific import whatsoever. It does, however, have value as a Victorian museum piece, made about 1900 by the prominent London taxidermy studio of Rowland Ward Ltd. Such dodo models were quite popular with natural history museums around the world, and some remain on display today. (More than a few visitors, breezing past the labels, have come away thinking they saw the real thing.) The model is fabricated of chicken feathers, duck wings, and—for the tail—the curled feathers of egrets.

Elephant Bird Egg

Egg of *Aepyornis maximus* (ee-pee-OR-nis), acquired in 1913.[4]

"Well . . . you've heard of the Aepyornis?"

"Rather. Andrews was telling me of a new species he was working on only a month or so ago. Just before I sailed. They've got a thigh-bone, it seems, nearly a yard long. Monster the thing must have been!"

"I believe you," said the man with the scar. "It *was* a monster. Sindbad's roc was just a legend of 'em."

—H. G. Wells, "Aepyornis Island"[5]

T he *Aepyornis*—commonly called the elephant bird—produced the largest eggs of any animal known on earth. The eggs of this extinct bird were far larger than the largest known dinosaur eggs. Elephant bird eggs make ostrich eggs look puny.

The egg pictured here is 12.2 inches tall by 8.6 inches wide and holds almost 2.5 gallons. By volume, it is as large as about 180 large chicken eggs, or about 7 ostrich eggs. When fresh, an elephant bird egg would have weighed about twenty pounds.

Elephant birds were limited to the island of Madagascar. Several species are known to have existed, but only the largest, *Aepyornis maximus*, lived on the island concurrently with human beings. They became increasingly rare as their habitat—grasslands and open forests—declined under pressure from plantations and other settlements.

Huge and flightless, elephant birds became extinct around 1700. No illustration of them has survived. But based on its skeleton, the elephant bird resembled an enormous ostrich, measuring up to eleven feet tall and weighing some eleven hundred pounds. By comparison, a large ostrich is about eight feet tall and weighs about three hundred pounds.

The elephant bird may well have been the inspiration for a mythical bird called the roc, known from the tales of the *Thousand and One Nights*. In those stories, Sinbad the Sailor tells of his encounter with a gigantic bird whose young were fed on elephants. A fictional elephant bird also appears in *Horton Hatches the Egg*, by Dr. Seuss (Theodor Seuss Geisel), whose real-life father was curator of a Springfield, Massachusetts, zoo.

Whole intact elephant bird eggs like this one are extremely rare. In 1933, when the last survey was made, only nineteen were known in the world.

A Sirenian Silenced

Skeleton of *Hydrodamalis gigas* (hy-droh-DAM-a-lis), collected from remains in 1882; in background, head of African elephant, *Loxodonta africana*.[6]

The skull shown here belongs to an extinct marine mammal called Steller's sea cow. Toothless and vegetarian, this gentle animal was extirpated by hunters in the eighteenth century. Its surviving relatives in the order Sirenia—the dugongs and manatees—are all listed as vulnerable to extinction.

Steller's sea cow was first recorded in 1741 by the German naturalist Georg Wilhelm Steller. He accompanied the Danish sea captain Vitus Bering on a harrowing voyage aboard the *St. Peter*, a Russian ship assigned to map the coast of Alaska. Following violent storms, their ship was beached on an unknown island. Bering died there—and the island was later named after him. The crew members, weak and sick with scurvy, made a heroic effort to rebuild their ship and sail back home. Many owed their survival to the sea cows they found grazing on seaweed off the island's coast.

Steller devotes an entire chapter to the sea cow in his riveting journal, published as *Journal of a Voyage with Bering, 1741–1742*. He describes the hungry sailors' strategy for hunting the enormous animals, which weighed roughly eight thousand pounds. A crew quietly rowed out to a herd of sea cows in a yawl, or small boat, carrying a harpoon attached to a long rope. When the harpooner struck a sea cow, men aboard the yawl thrust knives deep into its flesh. Finally, forty men on shore, holding the other end of the rope, hauled the dying sea cow to shore.

During the hunts, Steller noted, the sea cows demonstrated "an extraordinary love for one another."[7] When one was cut into, the others would try to overturn the yawl or form a circle around the injured sea cow to keep it from being hauled to shore. Sometimes they placed their bodies on the rope in an effort to pull out the harpoon, succeeding more than once.

Sea cow meat tasted like beef, Steller wrote, and it was "evident that all who ate it felt that they increased notably in vigor and health." The sea cows were no match for hunters. No matter how many were killed or wounded, Steller wrote, the herd remained in one spot. This made the animals easy targets for the Russian sailors and traders who later returned to the Bering Island area, stockpiling Steller's sea cows for provisions, until none remained. The species became extinct by 1768, just twenty-seven years after its discovery.

Harvard's rare skeleton of Steller's sea cow is a composite, made up of bones collected in 1882 by Leonhard Stejneger on Bering Island.

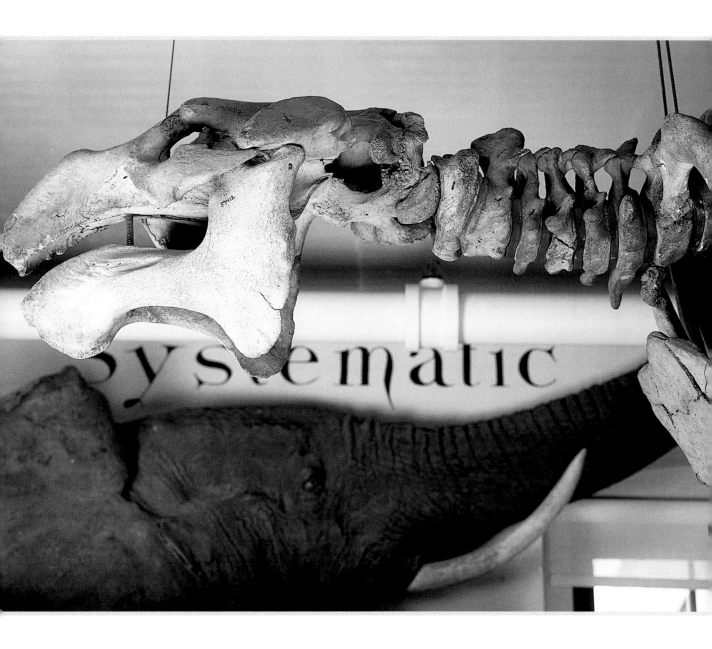

Of Auks and Postage Stamps

Mount of *Alca impennis* (AL-ca), collected before 1838; skeleton of same, collected from remains in 1887.[8]

The great auk received no special remarks in 1834 when Thomas Nuttall published his popular bird guide, *A Manual of the Ornithology of the United States and of Canada*. Although flightless, the great auk was a deft diver in icy seas. Thus, Nuttall wrote, "aided by all-bountiful Nature, it finds means to subsist, and triumphs over all the physical ills of its condition."[9]

About the triumph, Nuttall was sadly mistaken. An editor's note in the book's 1905 edition states:

> There is no mystery surrounding the extinction of these birds; they simply yielded to the inevitable law of the survival of the fittest. Through disuse the wings became unfit for service, and the parents could not reach a place of safety for their eggs; and though expert divers, and strong, swift swimmers, their legs were almost useless when upon land, and the birds were continually surprised by hunters and captured in large numbers, until the last one perished.[10]

The great auk had once been common in the North Atlantic, from Canada's Gulf of St. Lawrence and as far east as Norway. But the birds were terribly vulnerable, nesting in large colonies on land. Slow on their feet, great auks made easy prey for whalers and other hungry seafarers. They were also hunted for feathers. And they were killed for their oil, which was used—on Funk Island, off the Newfoundland coast—to fuel huge sailors' bonfires that burned for weeks. The last recorded sighting of a great auk was June 3, 1844, on the island of Eldey, off Iceland. The two birds, possibly a mated pair, were clubbed to death by fishermen.

The great auk mount shown here, one of about eighty stuffed auks known in the world, is missing a few bones. The mount once belonged to Rowland Hill, the English viscount who invented the postage stamp. Hill purchased his great auk in 1838 and had it remounted by a taxidermist named Henry Shaw. During the process, Shaw removed a handful of bones from the bird's wings and upper legs. Years later, he sold them for £4 to Lord Walter Rothschild, owner of a vast ornithological collection. These bones—with scientifically valuable tissue still clinging—remain in Tring, England.

Eventually, Hill's entire collection was liquidated, and the great auk—minus the few bones—put up for sale in London by the Rowland Ward taxidermy firm. John Eliot Thayer of Massachusetts acquired the mount and presented it to Harvard's Museum of Comparative Zoology in 1931, along with the rest of his magnificent bird collection.

The great auk skeleton shown here is complete, from a single bird. As such, it is even rarer than the mounted specimen.

A Skink, Extinct

Macroscincus coctei (ma-croh-SKINK-us), collected in 1900.[11]

This lizard, the Cape Verde skink, is almost certainly extinct. Its range was apparently quite small, limited to a few islands in Cape Verde, an archipelago nation located off the coast of Senegal. In the 1990s, Italian biologists who searched the islands found no evidence of the skink's existence.

The Cape Verde skink was once popular as a pet, and several herpetologists (experts on reptiles and amphibians) described its habits in captivity. Initially shy, the lizards gradually grew tamer, eating cherries, lettuce, and bananas from a keeper's hand. The skink had strong legs and claws and could hang upside down by its tail. Older individuals developed a thick, hanging dewlap—a fold of skin suspended from the throat. In the wild, the skink fed on mallow seeds and possibly bird eggs.

The specimen pictured here once lived in a zoo in Frankfurt, Germany. It was collected on the islet of Branco in 1900, along with other live Cape Verde skinks, for the Frankfurt Zoological Society. As the skinks died, they were sent to the Senckenberg Museum, Frankfurt's natural history museum. The last one arrived at that museum in 1913. It may well have been the last Cape Verde skink recorded anywhere.

This specimen came to Harvard in 1925 as part of an exchange with the Senckenberg Museum. In return, the Museum of Comparative Zoology sent several reptile specimens from Cuba, where Harvard had a botanical research station. The exchange was arranged by Thomas Barbour, the museum's curator of herpetology at the time and an authority on Cuban lizards. He requested the Cape Verde skink and six other specimens, explaining, "I want [them] very much for special study and have wanted them for years."[12]

Branco, the islet where this lizard was collected, is now uninhabited. Possibly, this is due to the terrible droughts that began afflicting Cape Verde starting in the mid-1700s. Three times, the resulting famines killed some 40 percent of the people living there. The recurring famines likely contributed to the decline of the Cape Verde skink, which was almost certainly hunted for food.

The Tasmanian Tiger, Continued?

Mount of *Thylacinus cynocephalus* (thy-la-SY-nus), acquired in 1882.[13]

Sometimes, extinct animals refuse to die. The Tasmanian tiger is a particularly good example of this phenomenon. The Tasmanian tiger, or thylacine (THY-la-seen), has likely been extinct since September 7, 1936, when the last known individual died in a zoo in Tasmania, an island of southeast Australia. And yet, residents continue to report mysterious sightings and signs of the animal, regularly publishing such accounts in local newspapers.

The Tasmanian tiger is not a tiger at all, but rather a marsupial resembling a dog with a striped back. It died out long ago on the Australian mainland, remaining only on Tasmania. During the 1800s, the government there placed a bounty on the animals' heads to appease angry sheep farmers. (In reality, sheep losses from thylacines were few, with wild dogs the true culprits.) Year after year, thylacines were shot, poisoned, snared, gassed, and trapped. All told, it took little more than a century of European settlement on Tasmania for the thylacine to become extinct.

Recently, biologists at the Australian Museum in Sydney have launched an ambitious—and probably impossible—effort to bring the thylacine back to life. Their plan is to clone the DNA of museum specimens. In 2002, they successfully replicated individual thylacine genes, by use of the polymerase chain reaction. The DNA they copied included that from a female pup preserved in alcohol in 1866. The scientists—led by museum director Michael Archer and geneticists Donald Colgan and Karen Firestone—showed that the specimens' DNA had not completely deteriorated over time.

The next step is to obtain sufficient thylacine DNA, using genomic amplification, so that a library of thylacine genetic material can be made (and stored) in bacteria. The biologists would then undertake large-scale sequencing of thylacine DNA, to make a genetic map of the species.

If this is successful, the scientists would attempt to make synthetic thylacine chromosomes and fertilize host cells, using technology that has not yet been invented. Their plan envisions reconstructing a complete set of thylacine genes in living cells inside a test tube. The nuclei of these cells could then be used to fertilize eggs from a host species that had been stripped of their original genetic material. In this way, a living animal could—in theory—develop. The Australian Museum's ultimate goal is to restore the Tasmanian tiger to the wild, as a viable breeding population. Most scientists remain highly skeptical about the project's viability.

Return of the Dawn Redwood

The discovery of this tree was one of the twentieth century's most thrilling plant stories. Botanists believed the dawn redwood had become extinct 2 million years ago. But in the 1940s, Chinese scientists were amazed to discover the tree growing in eastern Sichuan Province. *Metasequoia glyptostroboides* became a famous example of a "living fossil," an ancient species that has survived nearly unchanged.

The story began in 1941, when Chinese forester T. Kan spotted an unusual tree growing in the tiny village of Modaoqi. Local residents called it *shui-shan,* meaning "water fir." In 1946, botany student Chi-ju Hsueh was sent to investigate. He walked seventy-two miles to reach the tree, which, in winter, did not even have leaves. Nonetheless the trip was a success. Hsueh recalled, "The tree was gigantic; no one could have climbed it. As I had no specific tools, I could only throw stones at it. When the branches fell from the tree, I found, to my great surprise, that there were many yellow male cones and some female cones on the leafless branches. I jumped with joy and excitement."[14]

Hsueh's professor sent specimen fragments to Hsen Hsu Hu, the first botanist from China to receive a Harvard doctorate. Hu, recognizing the tree's remarkable lineage, placed it in the genus *Metasequoia,* previously known only from fossils.

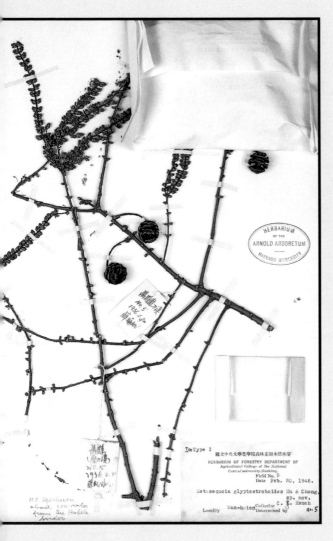

Hu then took action that forever changed the dawn redwood's prospects for survival. From Beijing, he sent word of his discovery to Elmer D. Merrill, director of Harvard's Arnold Arboretum. Merrill underwrote a seed-collecting expedition, from which he received more than two pounds of dawn redwood seeds. These he distributed to botanical gardens and arboretums across North America and Europe. From the seeds, huge trees have grown, some now standing more than a hundred feet tall.

Dawn redwoods are magnificent, with delicate, flat needles and reddish, furrowed bark. While they can reach towering heights, they are nonetheless dwarfed by their distant cousins, the giant sequoias of California, which grow to an average of 250 feet. Fossils of *Metasequoia* indicate that the tree has changed little over the past 100 million years, since the age of the dinosaurs.

Sadly, the dawn redwood is now listed as critically endangered in the wild, with just a few small stands remaining in Sichuan, Hunan, and Hubei provinces. Although the Chinese government protects mature dawn redwoods from being cut down, their habitats remain unprotected. With rice being cultivated right up to the tree trunks, new seedlings cannot survive. The original "living fossils" may not endure much longer.

The Last Wolf Nose

Snout of *Canis lupus lycaon* (CAY-nis); skull (on left) and partial skeleton of same subspecies; jaw and skull (on right) of *Canis lupus beothucus*; pelt of same subspecies; all specimens collected in the nineteenth century.[16]

The snout pictured here is believed to be the last soft tissue remaining anywhere of the wolves that once roamed New England. Because the animals frightened human beings and killed livestock, an all-out campaign to destroy them started as far back as the 1600s. Wolves have been extinct in New England since about 1897.

The dried snout contains DNA and so is quite important to biologists. The wolf it came from was collected in Penobscot County, Maine, by Manly Hardy. Hardy poisoned the animal to collect the bounty, probably no later than the 1880s. He saved the snout as a curiosity.

Wolf classification is a matter of considerable controversy. Most scientists believe that the wolf of New England—the eastern wolf—was a subspecies of the gray wolf. Others disagree and consider the eastern wolf a species distinct from the gray wolf, but virtually identical to the red wolf. And there are other views, too. Although these arguments may sound arcane, they are important in determining which wolf, if any, should be reintroduced in New England.

The photograph also shows bones of a Newfoundland wolf, a Canadian subspecies that became extinct by about 1930. The jaw and the right-hand skull belong to this subspecies, which existed only in Newfoundland, Canada. The speci-

mens were collected in 1865 by a trapper named J. M. Nelson, who sent to Harvard two complete skeletons. These are the only complete skeletons of the Newfoundland wolf known to exist.

The white pelt pictured here caused considerable confusion about the Newfoundland wolf's coloration. The Reverend Elwood Worcester shot the white wolf around 1896 and used its pelt as a rug, before presenting it to Harvard. For many years, this was the only known skin of a Newfoundland wolf, and so the subspecies was erroneously called the Newfoundland white wolf. Later studies showed that Newfoundland wolves existed in every shade, from pure black to pure white, and that they typically had the coloration of a German shepherd.

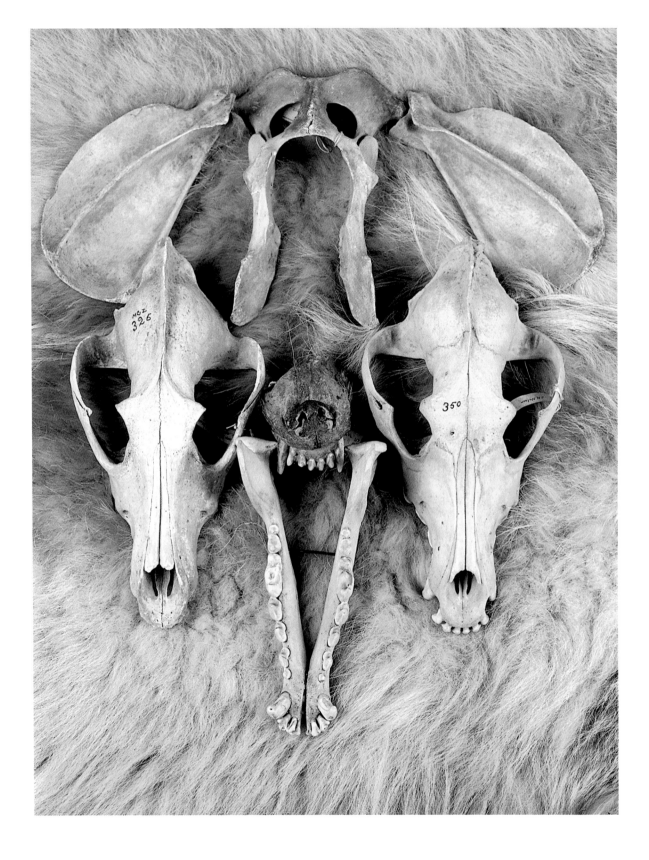

Fallen Butterfly

Glaucopsyche xerces (glaw-koh-SY-kee), collected before 1941.[17]

These butterflies are Xerces blues, the first U.S. butterfly species known to have become extinct due to human disturbance. The Xerces (ZERK-sees) blue had a small range, limited to California's San Francisco Peninsula. Urbanization overtook most of the sandy, open areas where the butterfly once lived.

Entomology professor W. H. Lange collected the last known Xerces blue on March 23, 1941, at San Francisco's Presidio military base. Lange knew that the species had grown scarce, but he believed small populations still remained at various sites near the city.

Years later, at age eighty-six, Lange returned to the spot on the Presidio where he had collected that last specimen, accompanied by science writer Mark Jerome Walters. Walters described the moment:

> Standing in an abandoned parking lot near the old naval hospital, [Lange] turned around a few times to get his bearings, mapping out old steps as he remembered them. "It would have been over there!" he finally announced, pointing with his finger down a slope and into a patch of underbrush. "I could never have imagined it would be the last seen alive."[18]

Xerces blues were small, lovely butterflies. In the photograph, the male appears at the bottom, with iridescent blue-violet wings. The female, with brown wings, is on the top. The middle specimen shows the spotted undersides.

No one knows the exact causes of the butterfly's disappearance. Possibly, the species' dwindling population lost the genetic diversity needed to withstand the stresses of urbanization. Also the Xerces blue, like many other butterflies in its family, might have had a cooperative relationship with particular species of ants. (See "Blue Butterflies and Ants," page 130.) When the Argentine ant was introduced from South America, those ties may have been fatally disturbed.

The Xerces blue was described in 1852 by French naturalist and physician Jean Alphonse Boisduval. He named the species for the Persian king Xerxes, using the French spelling. The butterfly is the namesake of the Xerces Society, an international organization dedicated to the conservation of invertebrates.

Disappearing Frogs

Rheobatrachus silus (ray-oh-ba-TRA-kus), collected in 1978.[19]

This amazing frog is gone forever. Known as the southern gastric brooding frog, it reproduced in a highly unusual way. The mother would swallow up to twenty-five fertilized eggs, which then developed in her stomach. During incubation, her entire digestive system shut down, and she ate nothing. After six or seven weeks, the fully formed froglets emerged from her mouth and hopped away.

Medical researchers were quite interested in gastric brooding frogs. They believed the frogs' ability to suppress the production of stomach acid could lead to treatments for people suffering from gastric ulcers. The two known species, however, became extinct before researchers could make much progress.

In 1973, when scientists first described the southern gastric brooding frog, it was quite abundant in the rain forests of Queensland, Australia. Less then a decade later, it had disappeared. The frog's extinction fits a disturbing pattern of global amphibian declines.

Historically, amphibians have been very successful life-forms. They first appeared at least 350 million years ago during the Devonian period, long before dinosaurs. Yet the changes wrought by human civilization may prove overwhelming. The die-offs have been so drastic that herpetologists around the world have formed a Declining Amphibian Populations Task Force. James Hanken, director of Harvard's Museum of Comparative Zoology, has been actively involved in the task force, which investigates the causes of amphibian declines and supports efforts to reverse the disturbing trends.

Die-offs among amphibians may be related to their thin, highly permeable skin. Biologists believe this leaves them vulnerable to a constellation of modern ills. Amphibians are, as a group, having trouble adapting to acid rain, water pollution, climate change, and increasing exposure to ultraviolet light due to the thinning of the ozone layer.

As if these were not enough, another scourge was discovered in the 1990s. Tree frogs at the National Zoo in Washington, D.C., suddenly began dying at a disturbing rate. Scientists confirmed that the deaths were caused by a fungus. At the same time, this chytrid (KI-trid) fungus was found to be causing amphibian die-offs in Central America and Australia. Although scientists are not sure what doomed the gastric brooding frog, the chytrid fungus is suspected. Chytrid fungus has since been linked to amphibian die-offs in many other parts of the world.

5 Stephen Jay Gould's Fast-Changing Snails
& Other Scientific Discoveries

Split the Lark—and you'll find
the Music—

—Emily Dickinson[1]

As one strolls through a traditional natural
history gallery, among the fossilized
bones and stuffed animals, cutting-edge
research is not what springs to mind. Yet museum
specimens are regularly sources of discovery. Scientists study them to gain insights into evolution,
ecology, conservation, taxonomy, population biology, and other disciplines—even medicine.

One dramatic example of the scientific importance of museum specimens concerned the insec-
ticide DDT. In the late 1960s, biologists began to
suspect that DDT was weakening the eggshells of
birds of prey, including bald eagles. How did they
prove their case? In part, by using eggs from
museum collections. Biologists measured the
thickness of eggshells collected long ago and compared them to the pesticide-weakened specimens.
Based on these and other studies, DDT was banned
in the United States in 1973.

This chapter focuses on discoveries made by
Harvard scientists who study animals, minerals,
and plants. Most of them use museum specimens

in their work. Andrew Knoll, an expert on the earliest life-forms on earth, studies rock samples
containing fossil microbes that lived hundreds of
millions of years ago. Ernst Mayr, the great evolutionary biologist, has written mainly on theoretical
matters. Yet even he used bird specimens in for-
mulating some of his early groundbreaking work
on species, and he has consistently upheld the
value of well-documented museum collections in
understanding evolution.

Occasionally, museum specimens contribute
to medical discoveries. When researchers found
promising anti-AIDS properties in a rain-forest
tree from Borneo, they returned to the original site
to collect more specimens. But the original *Calophyllum* tree had been cut down, and other similar
trees contained far less of the desired compound.
A Harvard botanist came to the rescue, providing
the precise identification that led researchers
to find the proper tree in the Singapore Botanic
Garden.

Museum specimens are critical to many other
research projects as well, at Harvard and beyond.
Scholars continue to name and describe new
species for science, from arachnids to minerals
to orchids. Also, scientists extract DNA from
museum specimens to gain insight into the evolution of species. Using DNA sequences, biologists

Szenicsite, a recent addition to the world's catalog of minerals. It was named in 1997 for Terry and Marissa Szenics, a husband-and-wife collecting team, who discovered the mineral at a small copper mine in northern Chile.

construct phylogenetic trees—similar to people's family trees—that reveal the ancestry of species and degrees of relatedness among kin. Scientists have used museum collections to investigate such topics as how fish fins evolved, how forests have changed over the past two thousand years, and why so many species of beetles exist.

In the future, scientists will undoubtedly continue to find new ways to tease information from museum specimens. A bird's nest is just a bird's nest, until an enterprising botanist discovers that the plants woven into it reveal trends in air pollution. Natural history collections are a storehouse for the scientists of today, and a cache of discoveries waiting to be made.

Ernst Mayr, Spanning a Century

FROM LEFT: Two specimens of *Melanocharis longicauda*, followed by *Crateroscelis nigrorufa*, *Timeliopsis fulvigula*, *Clytomyias insignis*, and *Melanocharis. longicauda*.[2] With books by Mayr, from his personal library.

The great evolutionary biologist Ernst Mayr obtained his Ph.D. in ornithology at the age of twenty-one. His next ambition was to study birds in the field. In 1927, Lord Walter Rothschild asked him to undertake a collecting expedition to an area of relatively unexplored mountains in New Guinea. With "the ambition and untiring enthusiasm of youth," Mayr agreed, leaving his native Germany at twenty-three to become, as he put it, "a tenderfoot explorer."[3]

In New Guinea, "every mountain range has its own endemic birds, so visiting mountains is very important," Mayr said. "You almost always find something new."[4] During two and a half years of arduous trekking, he collected some seven thousand bird skins for museums in England, Germany, and New York.

Among his takings were the birds pictured here, collected in 1929. All are endemic to New Guinea—meaning they exist nowhere else. The first two birds on the left are lemon-breasted berrypeckers, which inhabit midmountain forests, foraging for fruit and spiders. Also pictured, second from right, is an orange-crowned fairy wren, which skulks in thickets and bamboo.

As an up-and-coming scholar of South Seas birds, Mayr obtained a curatorship in 1932 at the prestigious American Museum of Natural History in New York. Ornithology still held his attention, but his thoughts were moving far beyond beaks and feathers. Comparing birds from the different mountain ranges of New Guinea, Mayr began drawing conclusions about the fundamental workings of evolution.

In 1942, he published his landmark book, *Systematics and the Origin of Species*, laying out his classic definition of species: "Species are groups of actually or potentially interbreeding natural populations, which are reproductively isolated from other such groups." A new species, he contended, arises when a population of a species becomes separated, often by geographical barriers. Eventually, such a population may evolve traits that discourage interbreeding with other populations. Not only do the animals usually begin to look different, but their DNA diverges as well.

After 1953, when Mayr came to Harvard, his focus shifted away from birds and became more theoretical, turning to the history and philosophy of evolutionary biology. There is no Nobel Prize in biology, but Mayr has won biology's "triple crown," the Balzan Prize, the International Prize for Biology, and the Crafoord Prize. He has continued to publish well into his nineties.

Lizard Look-Alikes

OVERLEAF: From left, on map of Caribbean: *Anolis angusticeps* (a-NOH-lis), collected in 1957; *Anolis insolitus*, collected in 1965; *Anolis occultus*, collected in 1971.[5]

Each of these anole lizards lives on a different island in the Caribbean. The three look almost identical and fill a similar ecological niche, spending most of their lives perching on twigs. And yet, surprisingly, they are not closely related. Scientists call such animals ecomorphs.

The term *ecomorph* was coined in 1972 by Harvard biologist Ernest E. Williams. He studied anoles of the Greater Antilles and found that they typically fit into six niches. On each island, he found anoles adapted to life in tree crown, twig, upper trunk, middle trunk, lower trunk, and bushes or grass. The similarities among any given group were striking. When he compared the "twig anoles" on different islands, for example, they were generally alike in color, size, body proportions, perch, foraging habits, and escape behavior.

The three specimens shown here are "twig anole" ecomorphs. All three are small anoles specially adapted to perching in the twigs of trees. At left is *Anolis angusticeps* of Cuba, and at center *Anolis insolitus* of the Dominican Republic. The specimen on the right is *Anolis occultus* of Puerto Rico. Despite their similarities, Williams contended that they were not close relatives, based on detailed study of their skeletons and scale patterns. Instead, he believed that evolution had produced remarkably similar anole communities on each island.

More recently, Williams's pioneering work has been confirmed through DNA comparisons. Biologists have constructed phylogenetic trees showing that the similar-looking anoles are indeed not closely related. Such trees indicate that members of ecomorphs on different islands probably last shared a common ancestor more than 10 million years ago.

Phylogenetic trees are like people's family trees, for they reveal ancestry and degrees of relatedness among kin. But unlike family trees, phylogenetic trees are constructed by comparing the sequences of molecules (bases) that make up DNA in different species. Molecular research on anole ecomorphs has been led by Jonathan Losos, a Harvard-trained evolutionary biologist who teaches at Washington University in St. Louis. More broadly, Losos has shown that members of the same anole ecomorph on different islands are virtually never close relatives.

For the record, Harvard's Museum of Comparative Zoology has the world's largest collection of lizards in the genus *Anolis*—some fifty thousand specimens.

Schultes' (Magic) Mushrooms

OPPOSITE: *Panaeolus campanulatus* Linnaeus variety *sphinctrinus* (pan-ee-OH-lus), collected in 1938.[6]

BELOW: Schultes' controversial Golden Guide.

In no small measure, these mushrooms paved the way for the psychedelic revolution. They were collected in Mexico by a young Richard Evans Schultes, who was trying to solve a great botanical mystery. According to Spanish conquistador accounts, the ancient Aztecs had used hallucinogenic mushrooms in their religious ceremonies. But for centuries, scientists had been unable to find and identify the fungus.

In 1938, Schultes (SHUL-teez) traveled to Oaxaca, Mexico, to conduct research for his doctoral thesis on plant uses by the Mazatec Indians. There, he learned of a sacred mushroom still being used for its vision-producing powers. During all-night ceremonies, traditional healers would invoke the mushroom's medicinal powers through prayer. It was, Schultes believed, the very same mushroom used by the Aztecs, one they called *teonanácatl*, "flesh of the gods."

Schultes brought back specimens to Harvard, where he identified the mushroom as *Panaeolus campanulatus*. His handwritten label, pictured here, states:

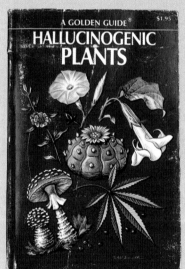

Said to be poisonous in overdoses of 50 or 60, but in moderate quantities produces hilarity and a general narcotic condition & well-being effect. Lasts for an hour or longer. Excess doses said to produce permanent insanity.

Schultes published an article about the mushroom in 1939 in Harvard's *Botanical Museum Leaflets*. Some fifteen years later, a copy came into the hands of R. Gordon Wasson, then a vice president of J. P. Morgan bank. Intrigued by Schultes' research, Wasson began making his own pilgrimages to Mexico in search of hallucinogenic mushrooms. In 1957, he wrote a famous article for *Life* magazine about his mind-altering experiences. A few years later, hippies by the hundreds—as well as Donovan and the Beatles—traveled to rural Mexico to try their own experiments with "magic mushrooms." (Neither Schultes nor Wasson employed the term.)

During his lifetime, Schultes published more than 350 books and papers, mainly in scientific journals. Among his writings for the general

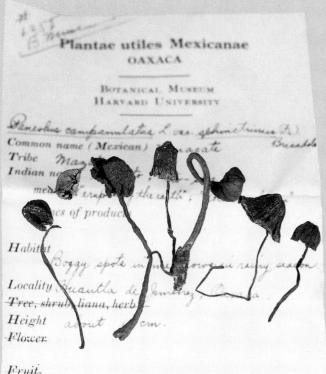

Plantae utiles Mexicanae

OAXACA

BOTANICAL MUSEUM
HARVARD UNIVERSITY

Panaeolus campanulatus L var *sphinctrinus* (F.)

Common name (Mexican) _____ _____ Breakable

Tribe Maz _____

Indian n_____

med_____ "crapo", the teeth",

_____ of produc_____

Habitat

Boggy spots in _____ in rainy season

Locality Huautla de _____, Oaxaca

~~Tree, shrub~~, liana, herb—

Height about _____ cm.

~~Flower~~

~~Fruit~~

Uses Said to be poisonous in overdoses of 50 or 60, but in moderate quantities produces hilarity and a general narcotic condition & well-being effect. Lasts for an hour or longer. Excess doses said to produce permanent insanity.

This is the teonanacatl of the ancient Aztecs. See Bot. Mus. Leafl., Harv. Univ., 7 #3, February 21, 1939.

Remarks Stem 1-2 mm. broad, cap 3 cm. broad and 1½ cm. high, hemispherical but often cuspidate, gills dark brown-black; plant coffee brown at base

Coll. No. 23_____ Coll. Richard Evans Schultes
27, July. ~~August~~. 1938   & Blas Pablo Reko

public, one little paperback is particularly loved by book collectors, *Hallucinogenic Plants: A Golden Guide*. Published in 1976, this was the most controversial Golden Guide ever written. It went through four printings before being withdrawn forever.

But Schultes' real fame rests on more lasting discoveries. He was a pioneer in the field of ethnobotany, the scientific study of the use of plants by indigenous peoples. During his years in Amazonia, he documented the uses of more than sixteen hundred medicinal plants among more than a dozen Indian peoples. Schultes was also one of the first scientists to warn of the perils facing both the plants and the indigenous peoples of the Amazon rain forest.

Stephen Jay Gould's Snail Shells

Drawers containing shells of *Cerion* species (SAIR-ee-on), photographed in the office of the late Stephen Jay Gould.

Stephen Jay Gould, the evolutionary biologist and renowned essayist, may not have approved of this photograph. He believed his findings on *Cerion* snail evolution had already received an unjustifiable amount of attention from the media. This book draws attention to them yet again, for what they reveal about evolution and Gould himself.

Like all good naturalists, Gould had remarkable powers of observation. While exploring Great Inagua Island in the Bahamas, he thought he spotted an example of evolution made visible. Strewn on a mud flat, he found large, finger-shaped shells of an extinct land snail, *Cerion excelsior*. He also found smaller, rounder shells of a land snail commonly found on the island today, *Cerion rubicundum*. And, even more interestingly, he found snails in intermediate forms, with characteristics of both the extinct and modern species.

Gould hypothesized that *C. rubicundum* had invaded the island at some point in the past, first creating hybrids and then becoming dominant, until it was the only form remaining.[7] To prove his theory, Gould and chemist Glenn Goodfriend used carbon dating to date the various forms of shells. The results were conclusive enough to be published in the prestigious journal *Science*. Gould had, in a sense, caught the snails in the act of evolving.[8]

Gould was bemused when the story ended up on the front page of major newspapers. In his essay collection *The Lying Stones of Marrakech*, he called this "the miselevation of everyday good work." He wrote, "A scanning of any year's technical literature in evolutionary biology would yield numerous and well-documented cases of such measurable, small-scale evolutionary change—thus disproving the urban legend that evolution must always be too slow to observe in the geological microsecond of a human lifetime."[9] And yet, for reasons elaborated in his essay, only the rare case captures the public imagination.

Undeniably, some of the snail excitement was due to the fame of Gould himself. A paleontologist by training, he may be best remembered for his popular writings about science. For twenty-five years, he wrote a masterful monthly column for the journal *Natural History*. Those columns were collected into ten books, many of them best sellers, including *Hen's Teeth and Horse's Toes* and *The Panda's Thumb*.

Gould, who taught at Harvard for many years, was also a prominent evolutionary scientist. He was best known for his controversial theory of punctuated equilibrium, the idea that evolution occurred not as a gentle, gradual process but rather in quick bursts. Shortly before his death in 2002, he published the massive book he had worked on

for some twenty years, *The Structure of Evolutionary Theory*. Publication was all the sweeter given that he had been diagnosed with a deadly form of cancer in 1982. He made a full recovery and went on to complete his magnum opus, before an unrelated cancer struck him down at age sixty.

Blue Butterflies and Ants

Jalmenus evagoras (jal-MEE-nus), various dates.[10]

This lovely Australian butterfly, the imperial blue, interacts with ants in fascinating ways. Ants "milk" its caterpillars as if they were tiny cows. When an ant taps the caterpillar with its antennae, the caterpillar releases sugary drops from a gland on its back. The ant drinks up the secretions, which are highly nutritious. As many as twenty-five ants may cover a single caterpillar.

In return for this food, the ants protect the caterpillars from dangers such as predatory wasps and parasitic flies. This type of cooperative relationship, in which both parties benefit from interacting with each other, is called mutualism.

The caterpillars of imperial blues are also unusual because they sing. Their "songs" are actually vibrations. When amplified to make them audible, they sound like hisses and grunts. Studies by Harvard's Naomi Pierce have shown that certain of these songs facilitate communication with ants. The caterpillars produce hisses, for example, only in the first five minutes after being discovered by a worker ant. (Ants, although nearly deaf to airborne sounds, are quite sensitive to vibrations.)

Studying the interactions of blue butterflies and ants is the specialty of Pierce, a distinguished entomologist (insect expert) who oversees a busy research laboratory. She and her graduate students examine such questions as why some species of blue butterflies interact with many kinds of ants, while others associate with only one ant species.

Her lab also studies the evolutionary consequences of specialization. Imperial blue butterflies cannot survive just anywhere. Their habitat must support their attendant ants, as well as the food plants eaten by their caterpillars. Such specialization puts constraints on blue butterflies, making them prone to faster rates of evolution. The results may include speciation—the formation of new species—or extinction.

The demise of one species of blue butterfly, the Xerces blue, may have been linked to its dependence on ants. (See "Fallen Butterfly," page 116.) The Xerces blue had a limited range, near San Francisco, California. Early declines in its population coincided with the introduction of the invasive Argentine ant, *Linepithema humile*, that displaced many native Californian ant species. By the early 1940s, the Xerces blue had become extinct.

An AIDS-Fighting Plant?

Calophyllum lanigerum variety *austrocoriaceum* (kal-oh-FILL-um), collected in 1987.[11]

Rain-forest plants can be lifesavers. This one has an inspiring story. In 1987, John Burley went on an expedition to Sarawak, a Malaysian state on the island of Borneo. Burley, a botanist at Harvard's Arnold Arboretum, collected the specimen shown here from an inconspicuous *Calophyllum* tree.

Back in the United States, researchers at the National Cancer Institute found that an extract from the plant showed promise in fighting HIV, the virus that causes AIDS. They sent a team back to Sarawak, only to discover that the original tree had been cut down. Other nearby *Calophyllum* trees contained far less of the targeted compound.

For help, the researchers turned to biology professor Peter F. Stevens, then a botanist at the Arnold Arboretum. Stevens knew his *Calophyllum*, having published a seven-hundred-page study of the genus. He inspected the various samples and concluded that researchers on the second trip had gotten the wrong plant. The original specimen had been a *Calophyllum lanigerum* of the variety *austrocoriaceum*. The plant from the second expedition was a *Calophyllum teysmannii*.

Armed with a proper identification, researchers tracked down living plants of *Calophyllum lanigerum* in the Singapore Botanic Garden. Next, they isolated the active anti-AIDS compound, named calanolide A. At that point, the National Cancer Institute, which conducts only preliminary research, passed the rights to the compound to MediChem Research, a biopharmaceutical company near Chicago. To fund further research, MediChem formed an innovative partnership in 1996 with the state government of Sarawak.

In 1999, a spin-off company, Advanced Life Sciences, assumed MediChem's interest in the public-private partnership. So far, clinical trial results of calanolide A have looked encouraging. Phase I tests showed that the compound appeared to reduce viral loads and was tolerated with only mild side effects. Phase II clinical trials are scheduled to begin in 2004.

Two New Fishes

Left, *Foetorepus goodenbeani* (fee-toh-REE-pus), collected in 1999; right, *Lepidopus altifrons* (le-pi-DOH-pus), collected in 1970.[12]

The world's oceans continue to yield surprises. Since 1985, scientists have discovered more than twenty new species of fish living in the waters off New England, despite these waters having been thoroughly studied for more than a century. Among the new species are the fish pictured here.

On the right are crested scabbardfish, first described and named in 1993. Little is known about these rare deep-sea fish, which can reach lengths of more than two feet. Although these specimens were collected in 1970, their identity remained a mystery for more than two decades.

On the left are palefin dragonets, a deep-sea species fairly common from Georges Bank to the northern Gulf of Mexico. Palefin dragonets live on sandy or muddy sea bottoms, usually at depths of more than six hundred feet. Alcohol storage has faded their colors, but in life the fish are red with olive brown spots.

Until recently, no one recognized that the palefin dragonet was new to science. Several American museums held specimens of the fish in their collections, but no one had carefully compared them. This sleuthing was conducted by two specialists, Tetsuji Nakabo of Kyoto University and Karsten Hartel, longtime fish collections manager at Harvard. They described and named the fish in 1999.

When scientists describe a new species, they do, quite literally, describe it. They examine every detail of its anatomy—its color, size, and form—and set down their findings in words. For their description to be made official, it must be published in a recognized journal.

In writing such descriptions, scientists use highly specialized terms that are all but incomprehensible to nonspecialists. To provide a little taste, here is a paragraph from Nakabo and Hartel's description of the male palefin dragonet:

> Body elongate and slightly depressed. Head slightly depressed. Eye large. Interorbital space narrow. Gill opening small, oval, located slightly posterior to the midpoint between dorsoposterior edge of eye and upper origin of pectoral fin. Preopercular spine without an antrorse process at base and with a forward process on the inner side; its posterior tip strongly curved upward.[13]

For enthusiasts of crossword puzzles and spelling bees, the word *antrorse* just might come in handy: it means directed forward or upward.

Ancient Life on Earth

Rock containing microfossils, collected in the 1980s.[14]

This rock may not look extraordinary, but it is key to scientists' emerging understanding of ancient life on earth. Viewed under a microscope, the rock reveals a wealth of fossil life-forms 800 million years old. These include bacteria, an amoeba, and early members of the green algae, the group that later gave rise to land plants.

Investigating life at its origins is the passion of Andrew Knoll, the eminent Harvard paleontologist. Among his specialties are cyanobacteria (sy-AN-oh-bacteria), the type of bacteria preserved in this rock. Cyanobacteria rank among the oldest life-forms ever found on earth. Some fossil specimens date back 2.7 billion years, and the group may have originated as much as a billion years earlier.

Cyanobacteria gave the planet its oxygen-rich atmosphere. Without them, humans and other oxygen-breathing organisms might never have evolved. Cyanobacteria were the first life-forms to conduct oxygenic photosynthesis, taking in carbon dioxide and water and releasing oxygen. Over several billion years, their activities transformed the earth's atmosphere, creating breathable air.

This rock, from Spitsbergen, an arctic island governed by Norway, was formed some 800 million years ago, when a storm washed tidal-flat sediments into a nearby lagoon. These sediments—rich in microorganisms—hardened into sedimentary rock. The rock's thin black stripes are the crucial parts, for they preserve the microfossils, some of the finest ever found.

Such specimens aid Knoll in his quest to explain the Cambrian Explosion, the sudden emergence of multicelled animals about 540 million years ago. To make sense of that explosion, he strives to understand what came before, at the very beginnings of life. He has done pioneering work on the fossils of the Proterozoic era, which began 2.5 billion years ago and ended some 2 billion years later. Knoll explains his research to nonscientists in his book *Life on a Young Planet: The First Three Billion Years of Evolution on Earth*.

As for cyanobacteria, they still exist today. Extremely hardy, they can live almost anywhere, with or without oxygen, even beneath the surface of rocks in the desert. Remarkably, over a span of 2 billion years, they have remained virtually unchanged.

E. O. Wilson's Ants

Ants have been the lifelong passion of Harvard entomologist Edward O. Wilson, one of the world's most distinguished biologists. Over decades of research and writing, he has shown a remarkable ability to see both small and large. He gazes patiently at the movements of tiny ants, while at the same time alerting humanity to the biodiversity crisis facing the entire planet.

This drawer contains *Pheidole* ants from Harvard's collection of more than 1 million specimens—the largest ant collection in the world. Their labels say "Wilson" because he was the first to name and describe them for science. To date, Wilson has named 337 *Pheidole* species from the western hemisphere, more than half of all those known. His hefty book *Pheidole in the New World* describes the ants belonging to this one "hyperdiverse" genus.

Wilson has made many discoveries about the complex world of ants. "If human beings were not so impressed by size alone, they would consider an ant more wonderful than a rhinoceros," he has stated. In the 1950s and 1960s, Wilson proposed and helped prove the then radical idea that ant societies communicate almost entirely by chemical signals—what we now call pheromones. Later, his studies of ant societies suggested that genes play a role in the social behavior of all animals, including humans.

Deepening his insights into animal societies, he founded a new discipline: sociobiology. While his thoughts on animals found general acceptance, his ideas about the genetic basis of human behaviors sparked tremendous controversy. In 1978, during a meeting of the American Association for the Advancement of Science, an enraged demonstrator poured a pitcher of ice water on his head. Many of Wilson's findings have endured, however; indeed, they have helped transform the entire field of biology and our understanding of human nature.

More recently, Wilson has focused on the biodiversity crisis. His most important book may well be *The Diversity of Life* (1992), which warns that human beings are causing a sixth great spasm of extinction, the most devastating since the dinosaur extinction 65 million years ago. He defines a new environmental ethic, arguing that humanity has an obligation to save entire ecosystems, not just individual species. "There can be no purpose more enspiriting," he writes, "than to begin the age of restoration, reweaving the wondrous diversity of life that still surrounds us."[16]

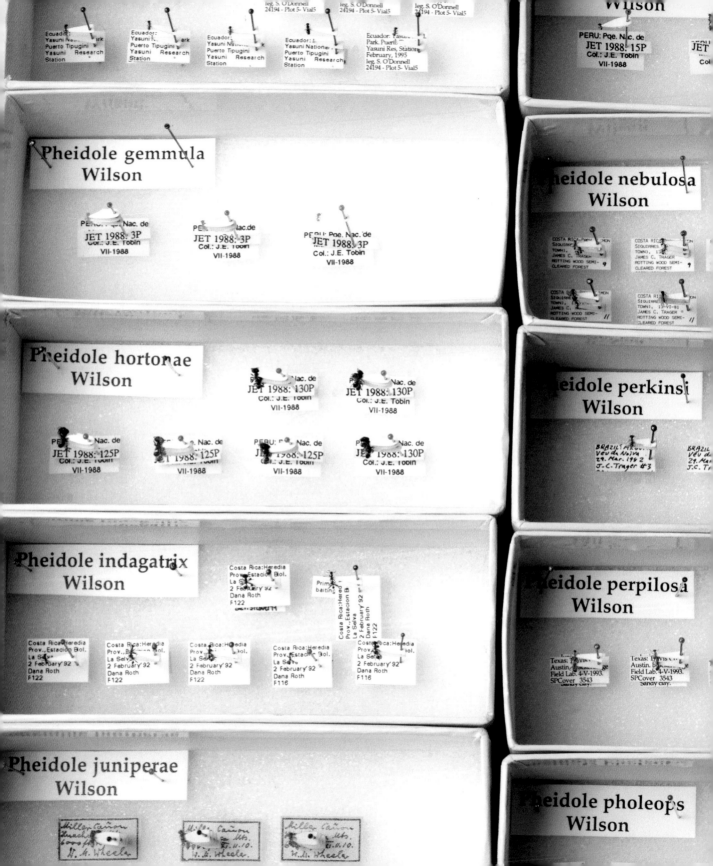

Pheidole gemmula
Wilson

Pheidole hortonae
Wilson

Pheidole indagatrix
Wilson

Pheidole juniperae
Wilson

Wilson

Pheidole nebulosa
Wilson

Pheidole perkinsi
Wilson

Pheidole perpilosa
Wilson

Pheidole pholeops
Wilson

J: Pqe. Nac. de
1988: 15P
.: J.E. Tobin
VII-1988

i
PERU: Pqe. Nac.de
JET 1988: 15P
Col.: J.E. Tobin
VII-1988

PERU: Pqe. Nac.de
JET 1988: 47P
Col.: J.E. Tobin
VII-1988

PERU: Pqe. Nac. de
JET 1988: 47P
Col.: J.E. Tobin
VII-1988

PERU: Pqe. Nac. de
JET 1988: 47P
Col.: J.E. Tobin
VII-1988

COSTA RICA: Prov.
SIQUIRRES,
TOWN),
JAMES C. TRAGER
ROTTING WOOD SEMI-
CLEARED FOREST
9

COSTA RICA: Prov. Lim
SIQUIRRES,
TOWN),
JAMES C.
ROTTING WOOD SEMI-
CLEARED FOREST
4

A RICA:
IRRES,
.C.
TING WOOD SEMI-
ARED FOREST
11

COSTA RICA: Prov. Lim
SIQUIRRES,
TOWN),
JAMES C.
ROTTING WOOD SEMI-
CLEARED FOREST
11

COSTA RICA:
Prov., TU
(CATIE) SENDERO
NATURAL. 8-VI-81
JAMES C. TRAGER
in rotting log

RICA:
SENDERO
.C.
TING LOG 17

COSTA RICA:
Prov., TURRI
(CATIE) SENDERO
NATURAL 8-VI-81
JAMES C. TRAGER
IN ROTTING LOG 17

COSTA RIC
Prov., TURRI
(CATIE) SENDERO
NATURAL 8-VI-81
JAMES C. TRAGER
IN ROTTING LOG 17

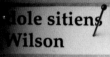

vic. 9-1
PERU. W.L. Brown
& W. Sherbrooke

vic.
PERU. W.L. Brown
& W. Sherbrooke

vic.
PERU. W.L. Brown
& W. Sherbrooke

rbrooke
PERU. W.L. Brown
& W. Sherbrooke

vic. 9-1
PERU. W.L. Brown
& W. Sherbrooke

Ti
vic
PERU. W.L. Brown
& W. Sherbrooke

L:
No. 3
Trager #3

BRAZIL: Mt.
Veu da Noiva
29. Mar. 1982
J.C. Trager #3

BRAZIL: Mt.
Veu da Noiva
29. Mar. 1982
J.C. Trager #3

r2.7mi SW
TX: Collings
13.7mi SW
14-X-1978. elev. 2300'
K.C. Neece. #21,773.
#21,773

TX: Colling
13.7mi SW
14-X-1978 elev. 2300'
K.C. Neece. #21,773.

TX: Colling
13.7mi SW
14-X-1978 v. 2300'
K.C. Neece. #21,773.

elev. 2300'
K.C. Neece. #21,773.

TX: Colling
13.7mi SW
14-X-1978. elev. 2300'
K.C. Neece. #21,773.

TX: Colling
13.7mi SW
14-X-1978. elev. 2300'
K.C. Neece. #21,773.

TX: Colling
13.7mi SW
14-X-1978. elev. 2300'
K.C. Neece. #21,773.

Texas: T
Austin. Brackenridge
Field Lab. 4-V-1993.
SPCover 3543
Sandy clay.

Texas: T
Austin. B
Field Lab. 4-V-1993.
SPCover 3543
Sandy clay.

in
stin. Brackenridge
Field Lab. 4-V-1993.
SPCover 3543

Texas: Tra
Austin. Br
Field Lab. 4-V-1993.
SPCover 3543
Sandy clay.

BO
JET 1988: 123B
Col J E Tobin
VII-1988

BO
JET 1988: 123B
Col J E Tobin
VII-1988

BO
JET 1988. 123B
Col J E Tobin
VII-1988

COLOMBIA: Chocó
1610m,
dGuaduales, by
Triver
1978. C. Kugler

COLOMBIA: Chocó
1610m, Finca
dGuaduales, by
Triver
1978. C. Kugler

Grenada: St. David's
Parish. La Sagesse
Bay. Natural
16-VI-1995 SPcover
elev. <10m. G-36

Grenada: St. David's
Parish. La S
Bay. Natural
16-VI-1995 SPcover
elev. <10m. G-36

Grenada: St. David's
Parish. La S
Bay. Natural cem
16-VI-1995 SPcover
elev. <10m. G-36

Worm from the Abyss

Riftia pachyptila (RIF-tee-ah), collected in 1985.[17]

Strange animals live in the deepest depths of the sea, and this giant tubeworm is one of the strangest. It has no mouth. It has no anus. For years, biologists wondered how it managed to survive.

The mystery was solved by Harvard biology professor Colleen Cavanaugh, in a eureka moment. During her first year of graduate school, Cavanaugh figured out the answer in the middle of class and jumped up to announce it. Bacteria! There had to be symbiotic bacteria living inside the tubeworms, producing their food for them.

The giant tubeworm shown here was collected by the submersible *Alvin* out of the Woods Hole Oceanographic Institution. The first giant tubeworms were discovered during an *Alvin* dive in 1977. Seen from the submersible's window, the tubeworms grow in spectacular clusters, appearing like "six-foot-long expletives, shouts of brilliance."[18] The long, white tubes of their bodies bear tips of brilliant red, filled with blood.

Giant tubeworms belong to a fascinating ecosystem: the hydrothermal vent community. Fed by volcanoes, warm waters rise from vents (cracks) in the ocean floor. Several hundred species of specially adapted animals live in these mineral-rich waters, including clams, limpets, tubeworms, and crabs. Some—including the clams and mussels—rely on symbiotic bacteria for their sustenance, just as the giant tubeworms do.

Inside the giant tubeworms, the symbiotic bacteria have evolved an unusual way of life. They metabolize sulfide, drawing on chemicals found in volcanic vent water. (This makes worm dissection a stinky undertaking, for sulfide smells like rotten eggs.) Just as plants derive energy from the sun through photosynthesis, the bacteria derive energy from sulfide—a process called chemosynthesis. Chemosynthesis makes it possible for giant tubeworms to survive without any light at all.

This way of life makes giant tubeworms highly resilient to global catastrophes. In his 1982 book, *The Fate of the Earth,* Jonathan Schell envisions a post-nuclear-war world, with nearly all plant and animal life destroyed. "At the outset," he posits, "the United States would be a republic of insects and grass."[19] But in the deepest depths of the sea, giant tubeworms—free from the need for sunlight—might well survive unscathed.

6 Vladimir Nabokov's Genitalia Cabinet
& Other Miscellany

The Museum at one time housed an unbelievable number of strange odds and ends accumulated through the years and saved because the old-time museum man thought it was a sin to throw anything out.

—Thomas Barbour[1]

After Barbour took over as director of the Museum of Comparative Zoology in 1927, he decided to clean house. He arranged for unwanted specimens to be left on the museum lawn, free for the taking. (Such drastic measures would never be allowed under modern museum practices.)

Yet despite Barbour's efforts, Harvard's natural history collections have preserved certain items of considerable fascination, albeit of limited scientific importance. Some of these appear in this chapter. Tapeworms saved from residents of upper-crust Boston neighborhoods may have scant scientific value, but they are intriguing bits of social history. Similarly, biologists may find little use for the cabinet in which Vladimir Nabokov stored the genitalia of dissected butterflies. But to lovers of literature, that same cabinet is distinctly thrilling.

In addition, this chapter features several specimens from the Mineralogical Museum, such as rare crystallized gold and a slice of a beautiful meteorite. These appear in this final chapter not because they are natural curiosities, but rather because mineralogical specimens, never having been alive, do not easily fit elsewhere in the book.

Also bearing mention are specimens that were considered for this chapter, though ultimately rejected for various reasons. These include:

1. THE STALIN ANT. This ant was present at a Kremlin dinner attended by Joseph Stalin. Among the other guests was Harlow Shapley, a Harvard astronomer and amateur entomologist, who preserved the specimen, a *Lasius niger,* by dropping it into his vodka.

2. A SHAMAN'S NECKLACE. This necklace—of the type worn by Siona Indian shamans in Colombia—is strung with vanilla beans, "horns" from rhinoceros beetles, seeds, shells of land snails, tobacco bundles, bird bones, cotton, glass beads, and the wings of splendor beetles.

3. INFLATED CATERPILLARS. These reveal the lost art of caterpillar inflation, a technique used to preserve specimens in the nineteenth century. After its innards were removed, the caterpillar was inflated and then mounted on a cork, permitting study of its markings and anatomy.

Such odds and ends may not lead to great scientific discoveries, but they enrich our sense of history and convey a kind of pure delight. Henry James once wrote that Sir John Soane's Museum of

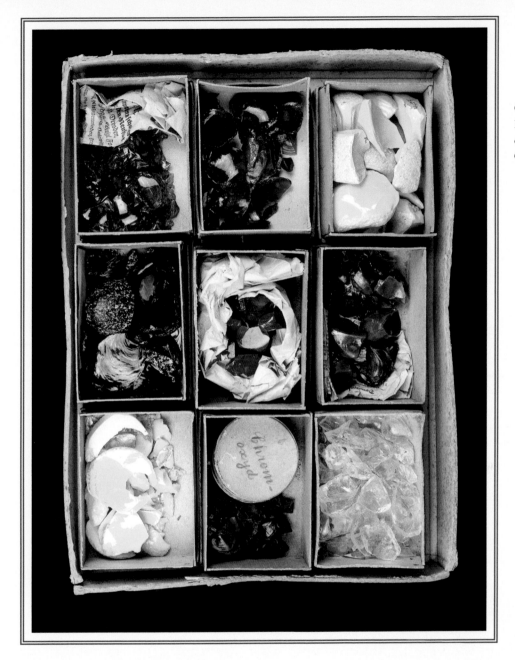

Cullet used by Rudolf Blaschka in creating his famous Glass Flowers.

art and architecture in London was "a very good place to find a thing you couldn't find anywhere else—it illustrated the prudent virtue of keeping."[2] The same certainly holds true for the natural history collections at Harvard.

Audubon's Untruth

John James Audubon was a brilliant painter and a complicated man. He claimed that he was born in Louisiana and even hinted that he had French royal blood. In fact, he was born in Haiti, the son of a French sea captain and his mistress.

Prior to Audubon, Alexander Wilson dominated the world of American bird books. Wilson, born a generation before Audubon, wrote and illustrated the pioneering nine-volume study *American Ornithology*. Audubon sought to eclipse Wilson, in part by being a superior artist, and also by claiming to be the first to depict various birds. As the painting shown here demonstrates, Audubon wanted priority for picturing the American ruffed grouse.

Audubon's grouse illustration, top right, catches him in a lie. He dated it June, 1805. The

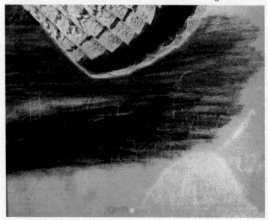

truth about this mixed-media work was known by 1931, when it was donated to Harvard by Gerrit Smith Miller Jr., then curator of mammals at the Smithsonian Institution. With the piece, Miller sent this delightful letter to Thomas Barbour, director of Harvard's Museum of Comparative Zoology:

> This morning I turned over for packing and shipping to you the Audubon drawing that I showed you the other day. The drawing was obtained by Greene Smith from J.G. Bell to whom it was given by Audubon himself. According to Bell's story, which I can remember hearing him tell, when I was a boy, Audubon and Wilson had a controversy as to which had been the first to depict our common grouse. To substantiate his claim to priority, Audubon submitted this signed and dated drawing. After the drawing came into his hands Bell discovered that the date written by Audubon was several years earlier than the date of the watermark on the paper, as may be seen by holding the picture up to the light. Hence much hilarity on the part of Bell and my uncle, which I can also remember.[3]

To make sure that Audubon's damning 1810 watermark could be seen by all who cared, the illustration was framed with glass on front and back. As for the John G. Bell mentioned in the let-

Audubon illustration of *Bonasa umbellus* (boh-NAY-sa), mixed media (chalk and watercolor). Wilson illustration of same, watercolor. (Photographs used by permission of the Ernst Mayr Library, Museum of Comparitive Zoology Archives, Harvard University.)

ter, he was a New York taxidermist who had accompanied Audubon on a trip up the Missouri River in the 1840s.

Miller thought the Audubon work belonged at the Museum of Comparative Zoology, together with the large bird collection of his great-uncle, Greene Smith. The museum also owned Wilson's watercolor of the ruffed grouse, shown here. (Wilson's painting is undated.) Barbour, the museum director, hung the competing grouse illustrations in his office, side by side.

A Cannibal Tale

Ornithoptera paradisea (or-ni-THOP-ter-ah), collected about 1900; with note from British butterfly broker.[4]

During the Victorian era, many a swashbuckling natural history collector headed for the tropics, searching for the beautiful and the rare. They did not always return. Some fell prey to snakebites, shipwrecks, or jaguars, others to diseases like yellow fever and malaria.

Around 1900, Carl von Hagen ventured to Papua New Guinea in search of butterflies. His was an unusual fate: he was eaten by cannibals. Oddly enough, his collection of spectacular birdwing butterflies survived.

One of his butterflies, a tailed birdwing, made its way to Harvard's Museum of Comparative Zoology. This handwritten note was attached, from an anonymous British broker:

> I sent you the one & only Orni[thoptera] paradisea it was ever my good fortune to receive, Carl v. Hagen who took this pair was afterwards eaten by the Papuans & the only thing he left his wife was about 4 pairs of these & I believe Staudinger secured them, at any rate when I wanted a pair for a customer a year or two ago I wrote to him & as you will see by enclosed he had some, but I thought £20 was too much to give to resell at a profit. I think you will agree you get them cheap for (virtually) £14-10-0.

The Staudinger mentioned above was Otto Staudinger, a prominent German butterfly collector, who described this species in 1893. The genus name *Ornithoptera* comes from the Greek words for bird (*ornithos*) and wing (*pteron*), referring to the unusual size and strength of these butterflies.

Some 30 species of birdwings exist, all of them from the Old World tropics. Caterpillars of birdwing butterflies are quite particular about their food, eating only a few species of tropical vine commonly called Dutchman's pipe. Voracious caterpillars can defoliate an entire vine. If overcrowded, the larger birdwing caterpillars sometimes eat the smaller ones, themselves becoming cannibals.

Glass Flowers and Beyond

OVERLEAF: Blaschka models of jellyfish *Rhizostoma pulmo* (ry-zo-STOH-ma), acquired in the 1880s, and diseased apple, *Malus pumila* (MAY-lus), made in 1932.[5]

Millions of people have made the pilgrimage to Harvard's collection of Glass Flowers, an icon of natural history in America. Visitors find it hard to believe that such stunningly realistic models could be fashioned from glass. Among the models are bananas and irises, orchids and carnivorous pitcher plants, each crafted with astonishing attention to detail.

The Glass Flowers were created by just two men, Leopold Blaschka (BLOSH-kah) and his son, Rudolf, in their studio near Dresden, Germany. In the 1860s, they became known for their scientific models of marine invertebrates, sought by museums worldwide. One of their jellyfish is shown here, with its eight long mouth-arms. In real life, the dome of this large rhizostome jellyfish may reach three feet in diameter. The dome of the Blaschkas' delicate model would fit in the palm of your hand.

The glass sea animals caught the eye of George L. Goodale, the first director of Harvard's Botanical Museum. He convinced the Blaschkas to leave invertebrates behind in order to produce an exhibition of fine botanical models for students and museum visitors. Beginning in 1887, father and son worked together at a furious pace for nearly a decade. The collection they created—underwritten by Elizabeth C. Ware of Boston and her daughter, Mary Lee Ware—was named the Ware Collection of Blaschka Glass Models of Plants. All told, it contains more than four thousand models.

After Leopold Blaschka's death in 1895, Rudolf persevered another forty years, working alone. The apple shown here belongs to his final group of models, affectionately known as the "rotting fruit." These illustrate fungal diseases in fruits of the rose family, such as peaches, apples, and apricots. This apple model, made in 1932, depicts a juicy variety called Emperor Alexander. It shows the effects of apple scab, *Venturia inaequalis*, which causes lesions on the leaves and olive brown spots on the fruits.

The composition of the model is quite complex. Rudolf Blaschka formed the apple using glassblowing techniques. To shape the leaves, he softened glass pieces in a flame, a process known as flameworking. He then strung the components on wires like beads on a necklace and fused them. The colors came from enamel powders he had made himself, affixed to the glass by gentle heating.

The Hamlin Necklace

Necklace of tourmalines, made in the late 1800s.[6]

Late in the autumn of 1820, amateur mineralogists Elijah Hamlin and Ezekiel Holmes spent a day exploring the mountains of Paris, Maine. As they headed home, Hamlin stopped on a hillside to admire the extraordinary sunset. Then, as his son Augustus later wrote, "On turning to the eastward for an instant for a final look at the woods and mountains in his rear, a vivid gleam of green flashed from an object on the roots of a tree upturned by the wind, and caught his eye."[7]

It was a crystal. Thrilled, the two returned home with their find, intending to return early the next morning to look for more. But snow fell during the night, and the hills remained buried the entire winter. At last, after the spring melts, the two students were able to return to the site. They discovered several crystals exposed on a ledge, and more buried in the soil, in shades of yellow, red, white, and green. The crystals turned out to be tourmaline, a colored stone used in jewelry. The site, later named Mount Mica, became America's first gem tourmaline mine.

For years, mining on Mount Mica continued as a small-time affair. Then, in 1868, Elijah and Augustus Hamlin launched an ambitious blasting program that uncovered new pockets. They discovered beautiful gem-quality tourmalines in the rainbow of colors showcased in this necklace.

The Hamlin Necklace is considered one of the most historic pieces of American jewelry, in part because of the prominence of the Hamlin family. Hannibal Hamlin, Elijah Hamlin's brother, served as vice president under Abraham Lincoln; Augustus Hamlin was mayor of Bangor, Maine. The necklace's central stone, a green tourmaline weighing 34.25 carats, was cut from an unusually fine crystal found at Mount Mica in 1886. Augustus Hamlin described it as "perfect except for a little flaw in the bottom."

Scientists have long been drawn to tourmaline, which develops an electrical charge when heated or rubbed. In 1880, Jacques and Pierre Curie first reported that when tourmaline is pressed or squeezed, one end of the crystal develops a positive charge, while the other end develops a negative charge. This property makes the mineral highly useful in pressure gauges. In 1945, near Alamogordo, New Mexico, blast-pressure gauges made from tourmaline measured the shock waves from the world's first atomic explosion.

High-Society Tapeworms

Taenia species (TEE-nee-ah), collected in the 1880s.[8]

During the 1880s, Dr. James B. Cherry preserved these tapeworms from the intestinal tracts of his patients. The addresses on his labels reveal that the parasite infected Bostonians from some of the city's finest neighborhoods.

One tapeworm, he noted, came from a male living on Ivanhoe Street, near ritzy Tremont Street. Another—forty-five feet long!—came from Miss Lottie Fowler of Hayward Place, close to the Boston Common.

Two tapeworm species are common in humans, the pork tapeworm, *Taenia solium*, and the beef tapeworm, *Taenia saginata*. People become infected by eating raw or undercooked meat containing immature parasites. Inside the small intestine, the immature tapeworms grow into ribbonlike adults. Tapeworms can survive for years, sometimes causing abdominal pain or weight loss. Miss Fowler's parasite was undoubtedly a beef tapeworm, as that species reaches much greater lengths than the pork variety.

As for Dr. Cherry, rather little is known about him. In the 1880s, he was practicing medicine at 157 Shawmut Avenue in Boston. According to his death notice in the *Journal of the American Medical Association*, he was born in 1844 and was a veteran of the Civil War. The obituary states that he received a medical degree in 1868 from the University of Pennsylvania Department of Medicine. Dr. Cherry's name does not, however, appear in that university's alumni catalog. In the nineteenth century, it was quite common for doctors to practice without a medical degree.

There is no record of why Dr. Cherry donated the tapeworms to Harvard's Museum of Comparative Zoology.

Gold!

Gold "horn," mined in 1887; gold crystal embedded in quartz, likely mined in the late nineteenth or early twentieth century; antique safe no longer used for storing gold.[9]

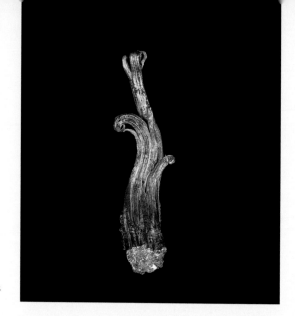

This gold "horn" is considered one of the most desirable mineral specimens in the world, valued for its uniqueness as well as its beauty. In exchange for this single specimen, a wealthy collector once offered to build Harvard a new mineralogy gallery.

Harvard has an unmatched collection of gold, due to the generosity of a private collector, Albert Cameron Burrage. In 1931, he bequeathed a fabulous collection of minerals to the Mineralogical Museum, including many examples of crystallized gold in exquisite forms. Crystallized gold—the rarest type—is prized by collectors. The delicate specimens must be extracted by hand. Crystallized gold is easily distinguished from typical gold nuggets, which look like mere lumps.

Knowledge of the gold horn's origins was lost until the 1990s, when a photograph of it was discovered in an obscure 1893 Rocky Mountain journal called *The Great Divide*.[10] The accompanying article states that the horn was found in 1887 in the Ground Hog mine of Eagle County, Colorado. By weight—about 8.5 ounces—it would have been worth $160 at the time. "But," says the article, "it has an increased value owing to its curious and rare formation." Over time, the specimen has continued to appreciate.

The lustrous specimen shown by the Diebold safe consists of gold crystals embedded in milky quartz. The Mineralogical Museum does not actually store its gold in this antique safe. For security reasons, the specimens are kept in a bank vault beneath Harvard Square.

Burrage, the great gold collector, graduated from Harvard in 1883. He served as a lawyer during the 1890s for the Brookline Gas Light Company during the "gas war" over Boston's utilities. When his client won the case, Burrage received some $700,000 in fees, thus establishing his fortune. He then turned to copper mining, organizing both the Amalgamated Copper Company and the Chile Copper Company. In his spare time, Burrage pursued two passions: tropical orchids and minerals.

Initially, his mineral collection served primarily as decoration for the billiards room of Burrage House, his French-style mansion, located on Commonwealth Avenue in Boston's Back Bay. Ultimately, his mineral collection grew so large that the billiard tables were displaced.

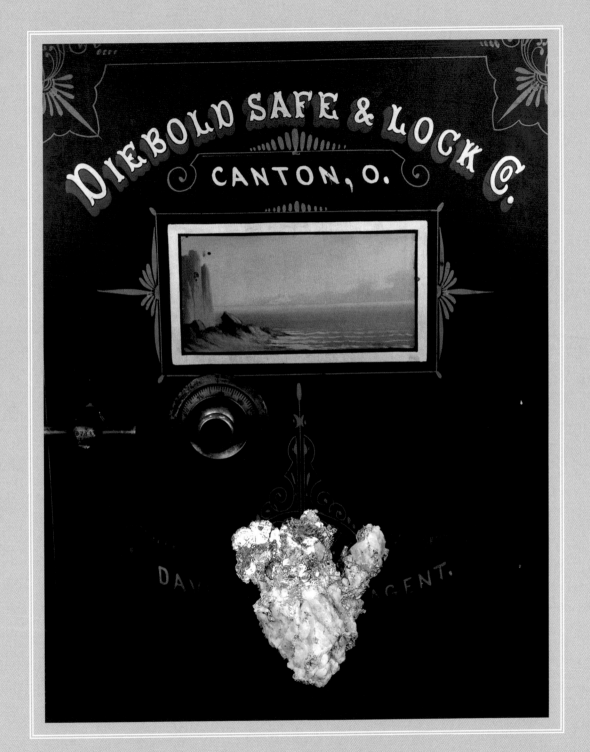

A Passion for Orchids

Images of *Pleurothallis lewisae* (pleur-oh-THAL-is), from about 1931.[11]

Oakes Ames found his lifework while still a teenager. He walked into the bedroom of his father, an heir to the Ames Shovel fortune, and saw two exquisite *Dendrobium* flowers on the bedside table. "There and then," Ames related, "I fell in love with orchids."[12]

Born into privilege, he was free to pursue his passion. As a Harvard undergraduate, he studied botany and toured the world's great orchid collections. In 1898, shortly after graduating, he was appointed assistant in botany, his first position at Harvard. There he would spend the next fifty years, as teacher, administrator, and scholar.

Over his lifetime, Ames described more than a thousand new species of orchids, including the one illustrated here. He named this one *Pleurothallis lewisae*, after Margaret Ward Lewis, who sent him the specimen from Guatemala. Like many orchids, this one is an epiphyte, meaning it grows on another plant. Epiphytic orchids do not harm their host, but merely cling on, absorbing moisture and dilute minerals that trickle past their roots. This one was found on a mango tree.

The top half of the page shows Ames's initial sketches of this *Pleurothallis* species, likely done as he peered through a microscope. At the bottom is the fine illustration that his wife, Blanche Ames, published in 1933. Blanche and Oakes Ames were close collaborators; she illustrated nearly all of her husband's scientific papers and books, including the seven-volume *Orchidaceae*.

A Smith College graduate, Blanche Ames was a suffragist and an outspoken advocate for women's right to birth control. And she was an inventor, who received her last patent at age ninety for an "anti-pollution toilet." Her political cartoons and some of her inventions can be seen at Borderland, the Ames's English-style stone mansion, now part of a state park in North Easton, Massachusetts. There, Oakes Ames kept his home botanical laboratory, with steel cases to hold his orchid specimens from all over the world.

That orchid collection, begun in his youth, took on considerable importance. Ultimately, he assembled an orchid herbarium of sixty-four thousand dried, pressed specimens, accompanied by photographs, drawings, and a large research library. In 1939, Oakes Ames donated the entire collection to Harvard, along with a generous endowment for its preservation. Summing up his long and satisfying career, he wrote simply, "Orchids dominated my life."

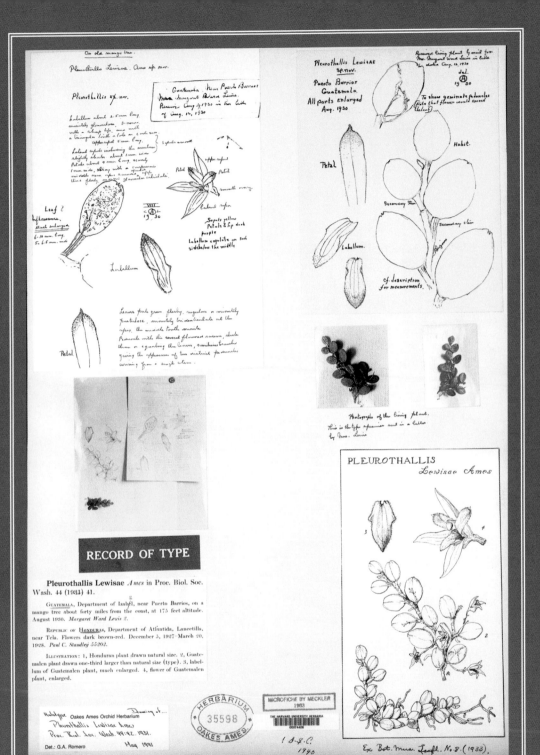

Extraterrestrial Gems

Pallasite, collected in 1951.[13]

Familiar meteorites look like chunks of rusty iron. But this is a slice of pallasite, the most beautiful of meteorites, shining and studded with translucent crystals. The yellow and green crystals of olivine, also called peridot, are lovely enough to be used in jewelry.

Nearly all meteorites are pieces of shattered asteroids that have fallen to earth. Scientists believe that pallasites come from deep within asteroids, where the metal core meets the rocky mantle. They likely formed as the asteroid—after becoming a molten ball—slowly cooled. Melted metal oozed up around the crystals in the overlying mantle, embedding them in a nickel-iron matrix.

Pallasites are named after Peter Simon Pallas, the German naturalist who was the first to describe them. During the late eighteenth century, he was invited by Catherine the Great to explore Siberia. In his travels, he came upon a large and unusual mass that had been found in the mountains near Krasnojarsk. Although he described the mass in great detail, he had no idea that it had come from space.

At the time, few scientists believed that extra-terrestrial objects could survive a fall to earth. Their doubts were finally overturned when a mete-orite shower dropped thousands of stony frag-ments on the town of L'Aigle in Normandy, France, on the afternoon of April 26, 1803. There were hundreds of witnesses, and an exhaustive investigation by French scientist Jean-Baptiste Biot proved that rocks do indeed fall from the sky.

Pallasites belong to a group called stony-iron meteorites, because they contain roughly equal amounts of metal and stony materials. Stony-irons are quite rare, comprising only about 1.5 percent of all known meteorite falls.

Meteorites are always named for the place where they were found. This mass was discovered

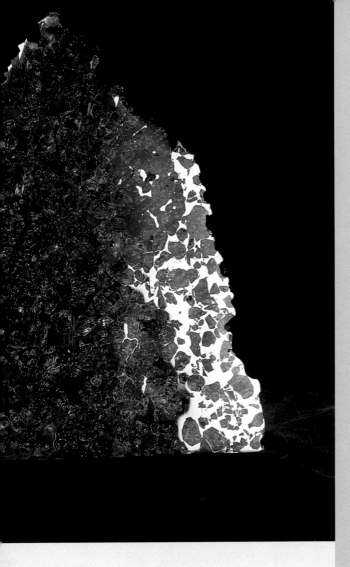

The Hide Room

The skins of lions and tigers and bears can all be found in the Hide Room. So can those of jaguars, cheetahs, zebras, wilde- beests, and thousands of the world's other large mammals. Many of the older hides were collected by scientists in the field. More recently, specimens have come from zoos, roadkill, state fish-and- wildlife offices, and U.S. Customs, which confiscates illegally imported animal skins. When this photograph was taken, the hides were hanging by hooks, a traditional storage method. Since then, they have been taken down and moved into mod- ern steel cabinets.

in 1951 by a rancher digging a hole in Esquel, Argentina, and so it is called the Esquel pallasite. The meteorite, weighing more than three thousand pounds, was later thinly sliced to allow light to shine through its crystals. This fine specimen was given to Harvard's Mineralogical Museum by Q. David Bowers, who donated his collection of 165 meteorites in the late 1990s.

Nabokov's Genitalia Cabinet

Cabinet used by Vladimir Nabokov for storing butterfly genitalia, from the 1940s.

Vladimir Nabokov, the great Russian-born writer, was also a gifted taxonomist. From 1942 to 1948, he worked as a research fellow at Harvard's Museum of Comparative Zoology, specializing in the butterflies known as blues. Mainly, he studied their genitalia.

Nabokov found genital structures far more useful than wings for classifying the blues. Often, two butterflies that appeared virtually identical to the naked eye proved to be quite distinct under the microscope. Indeed, Nabokov spent so many hours peering at butterfly genitalia—up to six hours a day, seven days a week—that his eyesight was permanently impaired. He focused mainly on the intricacies of the male genitalia, which he described as the "minuscule sculpturesque hooks, teeth, spurs, etc., visible only under the microscope."[14]

Nabokov stored his butterfly genitalia in the cabinet shown here. His specimens and notes have been preserved as he left them in the 1940s. Each glass vial, topped by a little cork, contains the genitalia of a single blue butterfly.

From his observations, Nabokov described several new species, often making divisions among butterflies that had previously been lumped together. In 1945, he named seven new genera of Latin American blues—an impressive feat, given that he had so few specimens at hand.

In 1948, Nabokov took a job teaching Russian literature at Cornell, and his career as a professional lepidopterist (butterfly specialist) came to an end. A decade later, the commercial success of *Lolita* freed him to write full-time. Butterflies appear often in his works. *Speak, Memory*, his autobiography, relates his boyhood passion for collecting butterflies on his family estate near St. Petersburg. And his early novel *The Gift* describes the butterfly-collecting expeditions of the narrator's lepidopterist father. In this passage, the father is introducing the narrator to his science:

> The sweetness of the lessons! On a warm evening he would take me to a certain small pond to watch the aspen hawk moth swing over the very water, dipping in it the tip of its body. He showed me how to prepare genital armatures to determine species which were externally indistinguishable.[15]

Nabokov wrote *The Gift* while living in Berlin, from 1935 to 1937. Perhaps not even he could have imagined that he would soon be at Harvard, with those genital structures a major focus of his life.

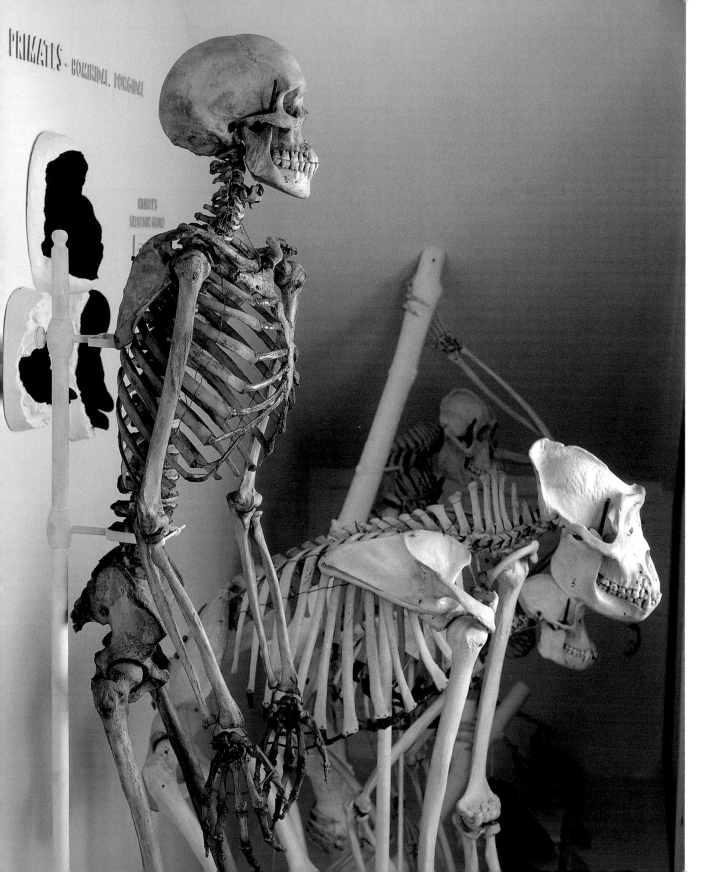

PRIMATES - HOMINIDAE, PONGIDAE

ORBITS
SCRATCHING GLAND

AUTHOR'S ACKNOWLEDGMENTS

As a French major in college, I learned nothing at all about natural history. I did, however, learn to appreciate the art of translation. In this book, I have tried to be a good translator, from a language called the scientific. It is the speakers of that difficult language who made this project possible. Scientists—at Harvard and around the world—kindly responded to my endless e-mails. They guided me, educated me, and corrected me, and to them I am deeply grateful. (Ultimately, of course, I alone bear responsibility for any information that was lost in translation.)

The concept of the book was initially shaped several years before I began working on it, by Jane Anderson Szele, formerly of the Harvard Museum of Natural History. Janis Sacco, the museum's director of exhibitions, then expanded the concept into a wonderful exhibition, Dodos, Trilobites & Meteorites...Treasures of Nature and Science at Harvard. Janis's work provided the foundation for this book, and her help throughout its shaping was invaluable.

Six Harvard scientists served on the book's scientific advisory group. They provided intellectual direction and expert advice, draft after draft, and they have my enduring gratitude. The members are Douglas Causey, Senior Vertebrate Biologist in the Museum of Comparative Zoology (MCZ); Carl Francis, Associate Curator in the Mineralogical Museum; Gonzalo Giribet, Assistant Professor of Biology and Assistant Curator of Invertebrates in the MCZ; James Hanken,

Alexander Agassiz Professor of Zoology, Curator of Herpetology, and Director of the MCZ; Karsten Hartel, Curatorial Associate of Ichthyology in the MCZ; and Donald Pfister, Asa Gray Professor of Systematic Botany, Curator of the Farlow Library and Herbarium, and Director of the Harvard University Herbaria.

For the chapter "Natural History at Harvard," Mary Winsor, author of Reading the Shape of Nature, the definitive history of the MCZ, generously took time to review the manuscript. Edward O. Wilson also read and commented on that chapter, and it was a privilege to have his input and encouragement.

Certain people at Harvard lent to this project not only their expertise but their enthusiasm, which meant very much to me. Judy Chupasko in mammals and José Rosado in herpetology were wonderfully kind. The entire Ornithology Department was extraordinarily helpful: Alison Pirie, Jeremiah Trimble, and (once again) Douglas Causey. I am also indebted to Adam Baldinger in malacology, Fred Collier in invertebrate paleontology, Ardis Johnston in invertebrates, and Phil Perkins in entomology. Emily Wood at the Harvard University Herbaria provided much assistance, as did Bob Cook at the Arnold Arboretum, who read many drafts.

At the Harvard Museum of Natural History, Tom Scanlon and Kyle Roberts provided critical aid with logistics and planning. Ed Haack and Bob Davidson also devoted considerable energy to the project. At the Ernst Mayr Library, Ronnie Broadfoot and Mary Sears

Primates, with skeleton of female human (*Homo sapiens*), received in 1883.

helped me countless times, always with good cheer, as did Dana Fisher in Special Collections. At the Harvard University Archives, Brian Sullivan and Michelle Gachette were particularly generous with their time.

In addition, I owe my thanks to scholars beyond Harvard—some halfway across the world—who responded to my queries. Here, I name only those experts not already cited in the text. For the historic holdings chapter, I received help on Captain Cook material from Jonathan King and Jenny Newell at the British Museum. Carolyn Gilman at the Missouri Historical Society provided assistance with Lewis's woodpecker. Gina Douglas at The Linnean Society of London helped with Linnaeus's dried, pressed fish. On Thoreau's specimens, invaluable aid came from Bradley Dean, formerly of the Thoreau Institute, as well as from Laura Dassow Walls at Lafayette College.

For the fossils chapter, vertebrate paleontologist Ralph Molnar, formerly of the Queensland Museum in Australia, initially provided help with *Kronosaurus*, but he went on to offer kind and patient guidance concerning many other specimens as well. Additional information on *Kronosaurus* came from Tony Thulborn at the University of Queensland and from Sue Turner at the Queensland Museum. On *Stupendemys*, I had assistance from Eugene Gaffney at the American Museum of Natural History. Regarding fossil insects, Michael Engel at the University of Kansas gently corrected my misunderstandings, and I also received help from Liz Brosius at the Kansas Geological Survey. Andrew Smith of The Natural History Museum, London, provided key information about Darwin's sand dollar; Ellis Yochelson of the Smithsonian Institution offered advice on Charles Walcott; and Catherine Forster of the State University of New York at Stony Brook provided help with *Triceratops*.

For the extinction chapter, Errol Fuller, a British expert on extinct birds, generously provided much assistance. On wolves, I was aided by John Maunder at the Provincial Museum of Newfoundland and Labrador; by Ronald Nowak, author of the most recent edition of *Walker's Mammals of the World*; and by Bradley White at Trent University, Canada. On the Cape Verde skink, I received help from Aaron Bauer at Villanova University, and from Gunther Köhler and Konrad Klemmer at the Senckenberg Museum in Germany. Michael Tyler at the University of Adelaide, Australia, read the piece on the gastric brooding frog. For the final chapter, Edward Burtt Jr. of Ohio Wesleyan University gave advice on Alexander Wilson, lepidopterist Kurt Johnson kindly checked over the Nabokov material, and Dr. Guy Kochvar assisted on tapeworms. For the "Natural History at Harvard" chapter, Alina Potts at Interlock Media tracked down a key photograph. Scholars who provided help with that chapter include molecular geneticist Andrew Mitchell at the University of Natal, South Africa; Tim White at the Peabody Museum of Natural History, Yale University; and environmental biologist Anne Yoder at Yale University.

In undertaking this project, Josh Basseches, friend since Amherst days, gave me the chance to pursue a dream. Molly Renda's book design was an inspiration. Greg Chaput at HarperCollins was a sharp-eyed editor. Anne Edelstein was not only an excellent agent but a kind and encouraging one.

Two friends versed in both science and literature, Rajesh Ranganathan and John Kleiner, bravely offered to read drafts and left them much improved. Bonnie O'Sullivan and Lisa Patterson provided the world's best child care, while I was away in Cambridge or sequestered in my study. My sons, Jacob and Milo, lights of my life, remembered all the facts that I forgot. And as for my husband, Lawrence Douglas . . . *without whom not.*

—N.P.

PHOTOGRAPHER'S ACKNOWLEDGMENTS

I would first like to thank Rosamond Purcell for paving the way and for ongoing inspiration. Next, I thank my photo assistant Margaret Soulman, who, despite her vegan philosophy, hoisted many a dead creature onto the platform in front of my camera. In addition, Margaret's calm demeanor and careful attention to detail made every photograph better.

The entire staff of HMNH was extremely cooperative and helpful. The curatorial staff of the Mineralogical Museum bore the brunt of my presence by allowing me to convert their collections area into a temporary studio for weeks at a time. Carl Francis and Bill Metropolis were patient, generous, and helpful. Ed Haack and Bob Davidson in exhibitions demonstrated consummate specimen handling skills as they wheeled various subjects into the portable studio. Thank you all.

I would also like to acknowledge support I received from the College of Charleston. The administration has been very supportive of my extracurricular activities since my arrival in 1994, and for this I am deeply grateful. Special thanks to Presidents Alex Sanders and Leo Higdon, Provosts Andy Abrams and Elise Jorgens, and Dean Valerie Morris. I am also indebted to the South Carolina Arts Commission and the Faculty Research Committee at the College of Charleston for their ongoing encouragement and support of my creative endeavors.

I offer special thanks to my wife, Michelle Van Parys, who, aside from making my participation in this project possible, provided critical feedback along the way. Thanks also to Mara Sloan, Andre Van Parys, Geoffrey Batchen, Roger Manley, Buff Ross, Anne Edelstein, and Molly Renda.

Technical notes: All of the color photographs in the book (with three noted exceptions) were made with a Sinar 4 x 5 view camera using natural light supplemented by daylight-balanced strobes. The film was Kodak Ektachrome EPN 100. There are no combined or digitally manipulated images.

—M.S.

PHOTO CREDITS FOR

NATURAL HISTORY AT HARVARD

All photographs in this chapter used by permission of
the Ernst Mayr Library, Museum of Comparative Zoology
Archives, Harvard University, except where noted below.

p.11 Courtesy of the Harvard University Archives

p.14 Courtesy of the Harvard University Archives

p.19 Courtesy of the Archives of the Gray Herbarium,
Harvard University

p.26 Courtesy of Interlock Media

NOTES

Natural History at Harvard

1. *Antony and Cleopatra*, 1.2.10–11.

2. Sara Schechner Genuth, "From Heaven's Alarm to Public Appeal," in *Science at Harvard University: Historical Perspectives*, ed. Clark A. Elliott and Margaret W. Rossiter (Bethlehem: Lehigh University Press, 1992), 28–29.

3. See *The Rarest of the Rare*, page 122.

4. Genuth, "From Heaven's Alarm," 29.

5. Samuel Eliot Morison, *Three Centuries of Harvard, 1636–1936* (Cambridge: The Belknap Press of Harvard University Press, 1963), 93.

6. I. Bernard Cohen, *Some Early Tools of American Science: An Account of the Early Scientific Instruments and Mineralogical and Biological Collections in Harvard University* (Cambridge: Harvard University Press, 1950), 39.

7. Francis Goelet, *The Voyages and Travels of Francis Goelet, 1746–1758*, ed. Kenneth Scott (Flushing, N.Y.: Queens College Press, 1970), n.p., Second Voyage, 25 October 1750.

8. David Murray, *Museums: Their History and Their Use*, vol. 1 (Glasgow: James MacLehose and Sons, 1904), 208. Stephen Jay Gould described cabinets of curiosity this way: "Collectors vied for the biggest, the most beautiful, the weirdest, and the most unusual. To stun, more than to order or to systematize, became

the watchword of this enterprise." Gould and Rosamond Wolff Purcell, *Finders, Keepers: Eight Collectors* (London: Pimlico, 1993), 17.

9. Broadside from 1764, "An Account of the Fire at *Harvard-College*," Cohen, *Some Early Tools*, fig. 2.

10. David Wheatland, *The Apparatus of Science at Harvard, 1765–1800* (Cambridge: Collection of Historical Scientific Instruments, Harvard University, 1968), 6.

11. Clifford Frondel, "The Geological Sciences at Harvard University from 1788 to 1850," *Earth Sciences History* 7, no. 1 (1988): 9.

12. Daniel Scott to Harvard Corporation, 14 June 1777, Corporation Papers, courtesy of Harvard University Archives.

13. It is possible that the tusks were from a hippopotamus, which was also sometimes known as a sea horse. The scientific name of the animal was not specified.

14. Minutes from "Meeting of the President of Fellows," 16 June 1777, Harvard Corporation Records, 2:465, Harvard University Archives.

15. Elizabeth Hall and Max Hall, *About the Exhibits*, 3rd ed. (Cambridge: Museum of Comparative Zoology, Harvard University, 1985), 41.

16. Ibid.

17. Charles Coleman Sellers, *Mr. Peale's Museum: Charles Willson Peale and the First Popular Museum of Nat-*

ural Science and Art (New York: W. W. Norton, 1980), 24.

18. John Woodforde, *The Strange Story of False Teeth* (New York: Universe Books, 1970), 51–52.

19. Broadside describing Waterhouse's natural history course, Cohen, *Some Early Tools*, fig. 10.

20. Frondel, "Geological Sciences at Harvard," 3.

21. John C. Greene, *American Science in the Age of Jefferson* (Ames: Iowa State University Press, 1984), 81.

22. James Johnston Abraham, *Lettsom: His Life, Times, Friends and Descendants* (London: William Heinemann, 1933), 480.

23. See letter from John White Webster to unnamed Harvard official, 20 May 1846, College Papers, 2nd ser., Harvard University Archives.

24. Cleveland Amory, "Dr. Parkman Takes a Walk," in *The Harvard Book: Selections from Three Centuries*, rev. ed., ed. William Bentinck-Smith (Cambridge: Harvard University Press, 1982), 132.

25. Agassiz did not coin the term *ice age*; Karl Schimper did. See Edmund Blair Bolles, *The Ice Finders* (Washington, D.C.: Counterpoint, 1999), 64–65. Agassiz did, however, do most of the work of convincing a skeptical scientific community that an ice age had existed.

26. Bernard Jaffe, *Men of Science in America* (New York: Simon & Schuster, 1958), 241.

27. Edward Lurie, "Louis Agassiz," in *American National Biography*, vol. 1, ed. John A. Garraty and Mark C. Carnes (New York: Oxford University Press, 1999), 176.

28. Mary P. Winsor, *Reading the Shape of Nature: Comparative Zoology at the Agassiz Museum* (Chicago: University of Chicago Press, 1991), 11.

29. John Himmelman, *Discovering Moths* (Camden, Maine: Down East Books, 2002), 115.

30. Samuel H. Scudder, "How Agassiz Taught Professor Scudder," *Every Saturday*, 4 April 1874; reprinted in *Louis Agassiz as a Teacher*, ed. Lane Cooper (Ithaca: Comstock, 1945), 57–58.

31. Longfellow wrote a poem in Agassiz's honor, "The Fiftieth Birthday of Agassiz," dated 1857. It imagines Nature as an old children's nurse, singing to Agassiz. "And whenever the way seemed long, / Or his heart began to fail, / She would sing a more wonderful song, / Or tell a more marvellous tale."

32. Henry Adams, *The Education of Henry Adams* (1907; repr., London: Penguin Books, 1995), 62.

33. A. Hunter Dupree, *Asa Gray: American Botanist, Friend of Darwin* (Cambridge: Harvard University Press, 1959; repr., Baltimore: Johns Hopkins University Press, 1988), 259.

34. Ibid., 330.

35. Richard Evans Schultes, "The Botanical Museum of Harvard University in Its 125th Year," *Botanical Museum Leaflets* 30, no. 1 (1984): 2.

36. Winsor, "Agassiz's Notions of a Museum: The Vision and the Myth," in *Cultures and Institutions of Natural History*, ed. Michael T. Ghiselin and Alan E. Leviton (San Francisco: California Academy of Sciences, 2000), 249–71.

37. Ibid., 251.

38. Thomas Barbour, *Naturalist at Large* (Boston: Little, Brown, 1943), 18.

39. Hall and Hall, *About the Exhibits*, 14.

40. Elizabeth Cary Agassiz, "The Hassler Glacier in the Straits of Magellan," *The Atlantic Monthly* 30 (1872): 472.

41. Elisabeth Deichmann, "Elizabeth Bangs Bryant," *Psyche* 65, no. 1 (1958): 5.

42. Richard H. Meadow et al., "In Memoriam" for Barbara Lawrence, www.ethnobiology.org/awards/lawrence/.

43. Edward O. Wilson, *Naturalist* (New York: Warner Books, 1995), 218–19.

44. Ibid., 220.

45. Bernd Herrmann and Susanne Hummel, *Ancient DNA* (New York: Springer-Verlag, 1994), 1.

46. Scott W. Rogers and Arnold J. Bendich, "Extraction of DNA from Milligram Amounts of Fresh, Herbarium and Mummified Plant Tissues," *Plant Molecular Biology* 5 (1985): 69–76.

47. Roger Vila Ujaldón, personal communication.

48. "Harvard Museum of Natural History Strategic Plan," unpublished, 10 January 2003.

Chapter 1: Meriwether Lewis's Last Bird & Other Historic Holdings

---

1. Charles Coleman Sellers, *Mr. Peale's Museum: Charles Willson Peale and the First Popular Museum of Natural Science and Art* (New York: W. W. Norton, 1980), 18.

2. *Drepanis pacifica* Gmelin, type specimen, MCZ no. 236875.

3. Thomas Barbour, *Naturalist at Large* (Boston: Little, Brown, 1943), 172–73.

4. Serpentine, Mineralogical Museum no. 54052.

5. Barites, Mineralogical Museum no. 64711, currently held by the Harvard Collection of Historical Scientific Instruments.

6. *Melanerpes lewis* Gray, type specimen, MCZ no. 67854.

7. Unfortunately, the labels were lost from nearly all the Peale specimens, making provenance difficult to confirm. As Harvard ornithologist Douglas Causey has pointed out, however, this specimen matches Alexander Wilson's illustration of Lewis's woodpecker so perfectly that most scholars are convinced of its authenticity.

8. *Cyclopterus lumpus* Linnaeus, MCZ no. 154782.

9. *Shortia galacifolia* Torrey & Gray, fragment of type specimen, HUH no. 00057718.

10. *Vaccinium oxycoccus* Linnaeus, HUH Botany Libraries archives, Henry David Thoreau Herbarium, Box 10, specimen 13.

11. Henry David Thoreau, *Wild Fruits*, ed. Bradley P. Dean (New York: W. W. Norton, 2000), 164–66.

12. *Conuropsis carolinensis* Linnaeus, MCZ no. 67853; hand-colored plate from *American Ornithology*, vol. 3 (1811), plate 26.

13. Alexander Wilson, *American Ornithology*, with continuation by Charles Lucian Bonaparte (London: Whittaker, Treacher & Arnot, 1832), 1:380.

14. Embryo of *Chelydra serpentina* Linnaeus, MCZ no. 138; *Contributions*

to the Natural History of the United States of America, plate IX-c. (Boston: Little, Brown, 1857).

15. Dallas Lore Sharp, "Turtle Eggs for Agassiz," *The Atlantic Monthly* 105, no. 2 (1910): 165.

16. *Rhabdopectella tintinnus* Schmidt, type specimen, MCZ no. 9213.

17. *Hoplostethus pacificus* Garman, MCZ no. 28765.

18. Barbour, *Naturalist at Large*, 139.

Chapter 2: Charles Darwin's Buried Treasure & Other Fossil Finds

---

1. Alfred S. Romer, "Fossil Collecting in the Texas Redbeds," *Harvard Alumni Bulletin* 42, no. 30 (1940): 1049.

2. *Iheringiella patagoniensis* Desor, type specimen, MCZ no. 102431.

3. Charles Darwin, *The Voyage of the Beagle*, The Harvard Classics, vol. 29 (New York: Collier and Son, 1909), 184.

4. Louis Agassiz, from handwritten note on his personal copy of *Catalogue raisonné des familles, des genres et des espèces de la classe des échinodermes*, quoted by permission of the Ernst Mayr Library, Museum of Comparative Zoology Archives, Harvard University. The translation is mine.

5. *Mammut americanum* Kerr, composite skeleton, MCZ no. 3410 with skull of MCZ no. 3411.

6. Idocrase, Mineralogical Museum collections.

7. *Isotelus gigas* DeKay, MCZ no. 100938.

8. *Prodryas persephone* Scudder, type specimen, MCZ collections.

9. Samuel H. Scudder, "An Account of Some Insects of Unusual Interest from the Tertiary Rocks of Colorado and Wyoming," *Bulletin of the United States Geological and Geographical Survey of the Territories* 4, no. 2 (1878): 524.

10. *Cycleryon propinquus* Schlotheim, MCZ no. 106092; *Plesioteuthis prisca* Rüppell, MCZ no. 104220.

11. *Kronosaurus queenslandicus* Longman, MCZ no. 1285.

12. *Dimetrodon milleri* Romer, MCZ no. 1365.

13. Romer, "Fossil Collecting," 1045.

14. *Meganeuropsis americana* Carpenter, type specimen, MCZ no. 4805.

15. *Stupendemys geographicus* Wood, MCZ no. 4376.

16. Skull of *Kayentatherium wellesi* Kermack, MCZ no. 8812.

17. Charles Schaff, personal communication.

Chapter 3: Seven-Colored Tanagers & Other Emblems of Biodiversity

---

1. Carl Linnaeus, *Systema Naturae*, 10th ed., 1758, as quoted in *The Book of Naturalists*, ed. William Beebe, (Princeton: Princeton University Press, 1988), 45.

2. *Lilium regale* Wilson, type specimen, HUH no. 00029975.

3. Ernest H. Wilson, excerpted from his 1928 book, *Plant Hunting*, reprinted in *The Flowering World of "Chinese" Wilson*, ed. Daniel J. Foley (London: Macmillan, 1969), 132.

4. *Xylophaga muraokai* Turner, type specimens, MCZ no. 316747; wood sample with bore holes, MCZ no. 168238.

5. Public Information Office,

Woods Hole Oceanographic Institution, "R.M.S. *Titanic* and Wood-Boring Mollusks: A Natural Progression," *The Titanic Commutator* 10, no. 2 (1986): 19.

6. *Hyridella fannyae* Johnson, MCZ no. 221348.

7. Richard I. Johnson, *The Nautilus* 62, no. 2 (1948): 48.

8. *Linophryne bicornis* Parr, MCZ no. 138063.

9. E. Bertelsen, "Notes on Linophyrnidae VIII," *Steenstrupia* 8, no. 3 (1982): 84.

10. Tim Birkhead, *Promiscuity: An Evolutionary History of Sperm Competition* (Cambridge: Harvard University Press, 2000), 45.

11. Top row, left to right: *Tangara nigroviridis*, MCZ no. 179255; *Chlorophanes spiza*, type specimen, MCZ no. 108199; *Poecilothraupis ignicrissa*, MCZ no. 266801; *Tangara fastuosa*, MCZ no. 28239; *Tangara mexicana*, MCZ no. 109919. Bottom row, left to right: *Tangara fastuosa*, MCZ no. 28238; *Euphonia musica*, MCZ no. 139455; *Tangara xanthogastra*, MCZ no. 139497; *Ramphocelus sanguinolentus*, MCZ no. 233682.

12. Skeleton of *Boa constrictor imperator* Daudin, MCZ no. R-176786.

13. Curculionidae spp., MCZ collections.

14. Mount of *Sphenodon* sp., MCZ herpetology collection.

15. *Dasypus novemcinctus* Linnaeus, MCZ no. 7271; *Chaetophractus nationi* Thomas, MCZ no. 7393; *Tolypeutes tricinctus* Linnaeus, MCZ no. 1975. Armadillo skin, *Euphractus sexcinctus* Wagler, MCZ no. 30949.

16. *Morpho* sp., MCZ collections.

17. Vladimir Nabokov, *Speak, Memory*, (New York: Knopf, 1999), 97.

1. Edward O. Wilson, "Biodiversity: Wildlife in Trouble," in *The Biodiversity Crisis: Losing What Counts* (New York: The New Press, 2001), 18.

2. Skeleton of *Raphus cucullatus* Linnaeus, MCZ no. 340825.

3. David Quammen, *The Song of the Dodo: Island Biogeography in an Age of Extinction* (New York: Simon & Schuster, 1997), 18.

4. Egg of *Aepyornis maximus* St. Hilaire, MCZ egg set no. 2251.

5. H. G. Wells, "Aepyornis Island," 1895, reprinted in *Strange Beasts and Unnatural Monsters: 13 Great Stories of the Macabre,* ed. Philip Van Doren Stern (Greenwich, Conn.: Fawcett Crest, 1968), 92.

6. Skeleton of *Hydrodamalis gigas* Zimmermann, MCZ no. 59412.

7. Georg Wilhelm Steller, *Journey of a Voyage with Bering, 1741–1742,* ed. O. W. Frost (Stanford: Stanford University Press, 1988), 162.

8. Mount of *Alca impennis* Linnaeus, MCZ no. 171764; skeleton of same, MCZ no. 340825.

9. Thomas Nuttall, *A Manual of the Ornithology of the United States and of Canada,* vol. 2, *The Water Birds* (Boston: Hilliard, Gray, 1834), 554. Nuttall taught natural history at Harvard from 1823 to 1833.

10. Nuttall, *Birds of the United States,* ed. Montague Chamberlain (Boston: Little, Brown, 1905), 416. As Harvard ornithologist Douglas Causey points out, Chamberlain's note reflects a common misconception of the period. His comment "Through disuse the wings became unfit for service . . ." reflects the enthusiasm at that time for Lamarckian explanations. Darwin, some forty-five years before this statement was written, proposed that structures like great auk wings arose by evolution through natural selection, and not by lack of use. Not until the mid-1940s did evolutionary concepts become well accepted throughout the scientific world—after the emergence of the New Evolutionary Synthesis, whose chief proponent was Harvard's Ernst Mayr.

11. *Macroscincus coctei* Duméril & Bibron, MCZ no. 21886.

12. Thomas Barbour to Robert Mertens, 17 August 1925, courtesy of the Senckenberg Museum archives.

13. Mount of *Thylacinus cynocephalus* Harris, MCZ no. 6349.

14. Chi-ju Hsueh, "Reminiscences of Collecting the Type Specimens of *Metasequoia glyptostroboides,*" *Arnoldia* 58/59, nos. 4 and 1 (1998–99): 9–10.

15. Fossil *Metasequoia* sp., HUH no. 56001; *Metasequoia glyptostroboides* Hu & Cheng, with foliage, HUH no. 2082; *M. glyptostroboides,* type specimen, HUH no. 00022512.

16. Snout of *Canis lupus lycaon* Schreber, MCZ no. 50518; skull and partial skeleton of same subspecies, MCZ no. 326; jaw and skull of *Canis lupus beothucus* Allen & Barbour, type specimen, MCZ no. 350; pelt of same subspecies, type specimen, MCZ no. 28726.

17. *Glaucopsyche xerces* Boisduval, MCZ collections.

18. Mark Jerome Walters, "Saying Goodbye," *National Wildlife* 37, no. 1 (1999): 38.

19. *Rheobatrachus silus* Liem, MCZ no. A-100691.

1. Emily Dickinson, "Split the Lark," *The Poems of Emily Dickinson,* ed. R. W. Franklin (Cambridge: Harvard University Press, 1999), 391.

2. From left: *Melanocharis longicauda,* MCZ no. 148146; *M. longicauda,* MCZ no. 148145; *Crateroscelis nigrorufa,* MCZ no. 154198; *Timeliopsis fulvigula,* MCZ no. 133020; *Clytomyias insignis,* MCZ no. 133021; *M. longicauda,* MCZ no. 133018.

3. Ernst Mayr, "A Tenderfoot Explorer in New Guinea," *Journal of the American Museum of Natural History* 32, no. 1 (1932): 83.

4. Mayr, personal communication.

5. From left, on map of Caribbean: *Anolis angusticeps* Hallowell, MCZ no. 59257; *Anolis insolitus* Williams & Rand, MCZ no. 83666; *Anolis occultus* Williams & Rivero, MCZ no. 128307.

6. *Panaeolus campanulatus* Linnaeus var. *sphinctrinus,* Economic Botany General Collection no. 6358.

7. William J. Cromie, "Snails Caught in Act of Evolution," *Harvard Gazette,* 9 January 1997, www.news.harvard.edu/gazette/1997/01.09/SnailsCaughtinA.html.

8. Harvard evolutionary biologist Richard K. Bambach has pointed out that what Gould showed was not evolution in the usual sense of one life-form evolving into another. Rather, he showed that an invading species survived and actually replaced the native species, which became extinct. The research was important because Gould was the first to document a "hybrid zone" in the fossil record, revealing intermediate forms

between a fossil species and a modern species. In addition, this was the first documentation of the length of time that hybridization could take place without one species simply "melding into" the other.

9. Stephen Jay Gould, *The Lying Stones of Marrakech* (New York: Three Rivers Press, 2000), 340.

10. *Jalmenus evagoras* Donovan, MCZ collections.

11. *Calophyllum lanigerum* Miquel var. *austrocoriaceum*, HUH no. 351.

12. Left, *Foetorepus goodenbeani* Nakabo & Hartel, MCZ no. 157182; right, *Lepidopus altifrons* Parin & Collette, MCZ no. 57374.

13. Tetsuji Nakabo and Karsten E. Hartel, "*Foetorepus goodenbeani*: A New Species of Dragonet *(Teleostei: Callionymidae)* from the Western North Atlantic Ocean," *Copeia* 1999 (1): 118.

14. Rock containing microfossils, HUH no. P-4353.

15. *Pheidole* ants: MCZ collections.

16. Edward O. Wilson, *The Diversity of Life* (New York: W. W. Norton, 1999), 351.

17. *Riftia pachyptila* Jones, MCZ no. 33230.

18. Cindy Lee Van Dover, *The Octopus's Garden* (Reading, Mass: Addison-Wesley, 1996), 76.

19. Jonathan Schell, *The Fate of the Earth* (New York: Avon Books, 1982), 65.

Chapter 6: Vladimir Nabokov's Genitalia Cabinet & Other Miscellany

---

1. Thomas Barbour, *Naturalist at Large* (Boston: Little, Brown, 1943), 168.

2. Henry James, *London Life* (1889; repr., New York: Grove Press, 1979), 85.

3. Gerrit S. Miller Jr. to Thomas Barbour, 23 October 1931, by permission of the Ernst Mayr Library, Museum of Comparative Zoology Archives, Harvard University. Miller's account of the argument between Audubon and Wilson is somewhat suspect. As Wilson scholar Edward H. Burtt Jr. has pointed out, Wilson died some fifteen years before Audubon began publishing. Burtt suggests that the ruffed grouse argument more likely occurred between Audubon and George Ord, the executor of Wilson's estate, whose dislike of Audubon was well known.

4. *Ornithoptera paradisea* Staudinger, MCZ collections.

5. Blaschka models no. 238, *Rhizostoma pulmo* Macri; no. 813, *Malus pumila* Miller.

6. Hamlin Necklace, Mineralogical Museum no. 2144.

7. Augustus Choate Hamlin, *The History of Mount Mica of Maine, U.S.A. and Its Wonderful Deposits of Matchless Tourmalines* (Bangor, Maine: privately printed, 1895), 1:8.

8. *Taenia* spp., MCZ nos. 37362, 37364, and 37389.

9. Gold "horn," Mineralogical Museum no. 98458; gold crystal embedded in quartz, Mineralogical Museum no. 110073.

10. "Gold in Spirals," *The Great Divide*, May 1893, 46.

11. Images of *Pleurothallis lewisae* Ames, HUH no. 35598.

12. Oakes Ames, *Orchids in Retrospect* (Cambridge: Botanical Museum, Harvard University, 1948), ix.

13. Pallasite, Mineralogical Museum no. 98.65.

14. Vladimir Nabokov to his sister, Elena Sikorski, quoted in *Nabokov's Blues*, by Kurt Johnson and Steve

Coates (Cambridge, Mass.: Zoland Books, 1999), 9.

15. Nabokov, *The Gift* (1963; repr., New York: Vintage International, 1991), 109.

INDEX

Page numbers in *italics* refer to illustrations.